GENETIC EFFECTS ON ENVIRONMENTAL VULNERABILITY TO DISEASE

The Novartis Foundation

The Novartis Foundation is an international scientific and educational charity (known until September 1997 as the Ciba Foundation), having been established in 1947 by the CIBA company of Basle, which merged with Sandoz in 1996 to form Novartis.

The Foundation promotes the study and general knowledge of science and encouraged international co-operation in scientific research. To this end, it organized internationally acclaimed meetings (typically eight symposia and allied open meetings and 15–20 discussion meetings each year) as well as publishing eight books per year featuring the presented papers and discussions from the symposia.

The Novartis Foundation's headquarters at 41 Portland Place, London W1B 1BN, provide library facilities, open to graduates in science and allied disciplines, and accommodation and meeting facilities to visiting scientists and their societies.

Towards the end of 2006, the Novartis Company undertook a review of the Foundation as a consequence of which the Foundation's Trustees were informed that Company support for the Foundation would cease with effect from the end of February 2008. The Foundation's Trustees decided that when the Foundation was dissolved, it would be taken over by the Academy of Medical Sciences. Accordingly, the Novartis Foundation and all its publications are now incorporated within **The Academy of Medical Sciences**.

The Academy of Medical Sciences

The Academy of Medical Sciences promotes advances in medical science and campaigns to ensure these are converted into healthcare benefits for society. Our Fellows are the UK's leading medical scientists from hospitals and general practice, academia, industry and the public service.

The Academy seeks to play a pivotal role in determining the future of medical science in the UK, and the benefits that society will enjoy in years to come. We champion the UK's strengths in medical science, promote careers and capacity buiilding, encourage the implementation of new ideas and solutions—often through novel partnerships—and help to remove barriers to progress. For further information visit *http://www.acmedsci.ac.uk*

The University of Otago
Dunedin, New Zealand

The University of Otago is New Zealand's oldest University. It was founded in Dunedin in 1869 and has earned a worldwide reputation for excellence, balancing the traditions of its history with modern scholarship, teaching and research.

Research underpins academic activity and professional training across the disciplines—Commerce, Health Sciences, Humanities and Sciences—with the result that the University of Otago is today New Zealand's most research-intensive university and also its top-ranked university for research excellence.

A commitment to internationalization, the aspirations of the indigenous Maori people, the fostering and commercialization of intellectual property, and contributing to the national good are other key areas of focus. The University has exchange agreements with some 90 institutions in 32 countries around the world.

The University of Otago has an academic presence that spans the length of New Zealand. Most undergraduate and postgraduate study takes place on the Dunedin campus, but the University is also represented in the cities of Auckland, Wellington, Christchurch and Invercargill. It has a roll of more than 20 000 students, with almost 4000 of these studying at the postgraduate level.

Further information can be found at
www.otago.ac.nz

Novartis Foundation Symposium

GENETIC EFFECTS ON ENVIRONMENTAL VULNERABILITY TO DISEASE

Edited by Michael Rutter
Institute of Psychiatry, King's College London, UK

John Wiley & Sons, Ltd

This publication is designed to provide accurate and authoritative information in regard to the subject matter covered. It is sold on the understanding that the Publisher is not engaged in rendering professional services. If professional advice or other expert assistance is required, the services of a competent professional should be sought.

Other Wiley Editorial Offices

John Wiley & Sons Inc., 111 River Street, Hoboken, NJ 07030, USA

Jossey-Bass, 989 Market Street, San Francisco, CA 94103-1741, USA

Wiley-VCH Verlag GmbH, Boschstr. 12, D-69469 Weinheim, Germany

John Wiley & Sons Australia Ltd, 33 Park Road, Milton, Queensland 4064, Australia

John Wiley & Sons (Asia) Pte Ltd, 2 Clementi Loop #02-01, Jin Xing Distripark, Singapore 129809

John Wiley & Sons Canada Ltd, 6045 Freemont Blvd, Mississauga, Ontario, Canada L5R 4J3

Wiley also publishes its books in a variety of electronic formats. Some content that appears in print may not be available in electronic books.

Genetic Effects on Environmental Vulnerability to Disease
x + 214 pages, 19 figures, 8 tables, 5 color plate figures

British Library Cataloguing in Publication Data

A catalogue record for this book is available from the British Library

ISBN 978-0-4707-7780-0

Typeset by 10½ on 12½ pt Garamond by SNP Best-set Typesetter Ltd., Hong Kong
Printed and bound in Great Britain by T.J. International Ltd, Padstow, Cornwall.

Contents

Participants

This book is the outcome of the Novartis Foundation symposium on Understanding how gene-environment interactions work to predict disorder: a lifecourse approach, held at The University of Central Otago, Dunedin, New Zealand, 8–9 November 2007

Editor: Michael Rutter

This symposium was based on a proposal made to the Foundation by Richie Poulton and David Skegg

Marco Battaglia Department of Neuropsychiatric Sciences, S. Raffaele University, 20 via d'Ancona, 20127 Milan, Italy

Antony Braithwaite Head of Cell Transformation Unit, Children's Medical Research Institute, University of Sydney, Australia

Kee-Seng Chia Department of Community, Occupational and Family Medicine, Yong Loo Lin School of Medicine, National University of Singapore, Blk MD3, Level 3, Medical Drive, Singapore 117597

Kenneth A. Dodge Center for Child and Family Policy, Terry Sanford Institute of Public Policy, Duke University, Box 90545 302 Towerview Drive, Durham, NC 27708-0545, USA

Andrew Heath Department of Psychiatry, Campus Box 8134, Washington University School of Medicine, St Louis, MO 63110, USA

Steven R. Kleeberger Environmental Genetics Group, National Institute of Environmental Health Sciences (NIEHS), Laboratory of Respiratory Biology, MD D2-01, Research Triangle Park, NC 27709, USA

Malak Kotb Mid-South Center for Biodefense and Security, University of Tennessee Health Science Center, VAMC 930 Madison Avenue, Memphis, TN 38163, USA

Nicholas G. Martin Genetic Epidemiology Group, Queensland Institute of Medical Research, PO Royal Brisbane Hospital, QLD 4029, Australia

Fernando D. Martinez Arizona Respiratory Center, 1501 N Campbell Ave., Rm 2349, P.O. Box 245030, Tucson, AZ 85724-5030, USA

Richie Poulton Dunedin Multidisciplinary Health and Development Research Unit, Department of Preventive and Social Medicine, Dunedin School of Medicine, PO Box 913, Dunedin, New Zealand

Anthony Reeve Cancer Genetics Laboratory, Biochemistry Department, University of Otago, 710 Cumberland Street, Dunedin, New Zealand

Stephen Robertson Clinical Genetics Group, Women's and Children's Health, School of Medicine, University of Otago, PO Box 913, PO Box 913, Dunedin, New Zealand

Michael Rutter *(Chair)* Box Number PO80, M.R.C. Social, Genetic & Developmental Psychiatry Centre, Institute of Psychiatry, King's College London, De Crespigny Park, Denmark Hill, London, SE5 8AF, UK

David Skegg Vice-Chancellor, University of Otago, PO Box 56, Dunedin, New Zealand

Harold Snieder Unit of Genetic Epidemiology & Bioinformatics, Department of Epidemiology, University Medical Center Groningen, University of Groningen, Hanzeplein 1 (9713 GZ), PO Box 30.001, 9700 RB Groningen, The Netherlands

Jim Stankovich Bioinformatics Division, Walter & Eliza Hall Institute of Medical Research, 1G Royal Parade, Parkville, Victoria 3050, Australia

Frédérique Tesson Director, Laboratory of Genetics of Cardiac Disease, University of Ottawa Heart Institute, 40 Ruskin Street, Ottawa, ON K1Y 4WY, Canada

Rudolf Uher Box Number PO80 Social, Genetic & Developmental Psychiatry Centre, Institute of Psychiatry, King's College London, De Crespigny Park, London SE5 8AF, UK

1. Introduction: whither gene–environment interactions?

Michael Rutter

GDP Centre, PO 80, Institute of Psychiatry, De Crespigny Park, Denmark Hill, London SE5 8AF, UK

As Tabery (2007) clearly outlined, there have been two entirely different concepts of gene–environment interaction (G × E). On the one hand, there was Ronald Fisher's biometric concept that treated G × E as a purely statistical feature. In effect, G × E was just a 'nuisance' term that needed to be removed in order to proceed with the serious business of partitioning the variance into genetic and environmental components. Fisher demonstrated the important effects of variations in scaling in both creating artifactual interactions and also as a means of getting rid of interactions. In addition, he noted the importance of taking gene–environment correlations into account when dealing with interactions. Fisher had a hugely influential role in developing statistical approaches to genetic issues (as well as very many other matters) and for many years his dismissal of G × E led behavioral geneticists to be similarly dismissive (Plomin et al 1988, Wachs & Plomin 1991).

Nevertheless, there are several reasons why we need to turn aside from the biometric concept of G × E as simply a statistical phenomenon. To begin with, Fisher's work in the field of population genetics was undertaken at a time when it could only be 'black box'. That meant that G × E had to be assessed in terms of an interaction between the totality of anonymous genes and the totality of anonymous environments. Approached in that way, it is scarcely surprising that very few interactions were found (Rutter & Pickles 1991). Rather, the biological expectation had to be in terms of specificities and not totalities.

Second, there was an exclusive focus on multiplicative interactions based on a logarithmic scale. Conceptually, the expectation is often of a simple interaction that is synergistic but not logarithmic (Greenland & Rothman 1998, Rutter 1983, 2006a, Rutter & Pickles 1991). The alternatives are most easily illustrated by reference to findings from a study by Brown and Harris that gave rise to controversy over whether or not there was an interaction between so-called vulnerability factors and provoking agents in the genesis of depression (Brown & Harris 1978). The base rate for depression in the absence of either was less than 2%; the provoking agent on its own increased the rate to 17% but the vulnerability factor on its own had a

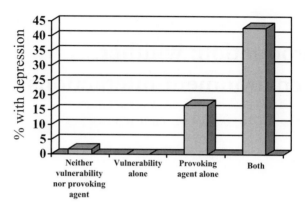

FIG. 1.   Interaction of vulnerability factors and provoking agents in the risk for depressive disorders in women (data from Brown & Harris 1978).

zero effect. On a simple additive model, the two in combination would be expected to lead to a rate of 17%, but the actual rate found was 43%—showing a statistically significant synergistic additive interaction (Fig. 1). There was not, however, a significant multiplicative interaction (between ratios, using a logarithmic scale). There is no satisfactory statistical way of deciding between these two varieties of synergistic interaction and there has to be reliance on what is hypothesized to be happening.

   Third, a statistical interaction can only be found if there is variation in both the genes and the environment. That means, for example, that there could be no statistical interaction in the case of phenylketonuria (because in all ordinary circumstances phenylalanines are a constituent of everyone's diets). This lack of a statistical interaction occurs despite the fact that the relevant mutated gene operates entirely through interaction with the environment. It might be thought that that is a rather unusual extreme example, but it is not. Thus, for example, it has been found that there is a genetic influence on susceptibility to malignant malaria. That will not result in a statistical interaction in the case of individuals living in an area where malaria is endemic. Similarly, there is a strong genetic component to hay fever but, once more, an interaction will not be found ordinarily because the pollens responsible for hay fever are widespread and differ little from individual to individual. We need to conclude that G × E is not just a statistical concept, although its detection involves crucially important statistical issues (see Greenland & Rothman 1998, Rutter & Pickles 1991).

   During the same period of time that Fisher was strongly arguing for the strictly biometric view of G × E, the eminent mathematician and biologist Lancelot Hogben argued equally strongly for a developmental notion of G × E that needed to be tackled through developmental biology rather than mathematical manipulations. Hogben's concept was most strongly taken forward by Waddington. For both of them, G × E was not a nuisance term to be removed by scaling modifications, nor was it an answer in its own right. Rather, as others have argued subsequently (Rutter & Pickles 1991), the finding of a G × E interaction points to a phenomenon that requires a biological understanding. That is the key issue that we will need to consider in our deliberations during the course of this symposium.

Should G × E be expected to be common?

In that connection, we need to consider whether there are reasons for expecting G × E interactions on the basis of what we know about how biology works (see Rutter 2006a). It is clear that there are several quite powerful reasons for expecting G × E to be a reasonably common phenomenon. First, evolutionary theory proposes that the mechanism for evolutionary change lies in genetically influenced variations in the adaptations of organisms to their environment. Note that it is not just that organisms vary in their response to environments but, more particularly, it is that their origin in <u>genetic</u> variations provides for evolutionary change.

Second, to suppose that all genes work by direct effects that are independent of the environment requires the proposition that sensitivity to the environment is the one and only biological factor that is uniquely outside genetic influence. A very large amount of data has shown that virtually all biological features involve substantial genetic influence. Why should sensitivity to the environment be the one characteristic that provides a major exception? It just does not seem a plausible supposition.

Third, a substantial body of research, both naturalistic and experimental, in a range of animal species, using a range of environmental manipulations, has shown huge individual variation in people's responses to both environmental hazards and environmental opportunities (Rutter 2006b). Given that such individual variation is so universal, is it plausible that genes play no role in this variation? That, too, seems most unlikely.

In addition, although perhaps less strongly persuasively, the reverse has also been shown. That is to say, animal studies of genetic effects have been found to vary substantially between experimental laboratories (Crabbe et al 1999, Mackay & Anholt 2007, Wahlsten et al 2006). That, too, suggests environmental moderation of genetic effects (i.e. the parallel of genetic moderation of environmental effects). The reason why this evidence is not quite so persuasive as it might be otherwise is that the laboratory variation, at least up to now, has not been tied down to particular laboratory conditions.

These considerations provide the backdrop to the issues to be considered during this symposium. I suggest that our starting point needs to involve both a focus on

G × E as a biological phenomenon, and not a statistical concept, and an expectation that G × E may well be rather more important, and more pervasive, than demonstrated so far. Nevertheless, the likely importance of G × E definitely does not mean that we should accept claims of its occurrence uncritically. To the contrary, we need to use rigorous research strategies, and thorough data analysis to put the claims to the test. What we will need to do is discuss how best to do this and what standards we need to employ. Some of the papers focus on conceptual or methodological issues and others will provide findings on different branches of medicine. The organizers of the symposium have deliberately opted to cover different medical conditions in order that we may consider the extent to which there are commonalities or differences in the approach to G × E when we move from one medical condition to another.

G × E and psychopathology

Let me use the example of G × E in the field of psychopathology as the peg on which to hang a set of suggestions of issues that I hope we can tackle. As we are meeting in Dunedin, it seems appropriate that I choose examples from Moffitt and Caspi and Poulton's research into G × E based on the Dunedin Longitudinal Study (Caspi & Moffitt 2006, Moffitt et al 2005, 2006, Rutter 2007a, Rutter et al 2006).

There are four published papers following the same basic strategy (Caspi et al 2002, 2003, 2005, 2007). For present purposes, I will focus on just the first two papers (Plates 1–3), but the same points arise similarly from the third and fourth papers.

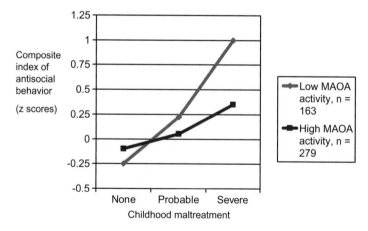

PLATE 1. Antisocial behavior as a function of MAOA activity and a childhood history of maltreatment (from Caspi et al 2002). A full-color version of this figure is available in the color plate section of this book.

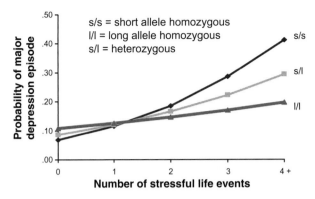

PLATE 2. Effect of life stress on depression moderated by *5HTT* gene (from Caspi et al 2003). A full-color version of this figure is available in the color plate section of this book.

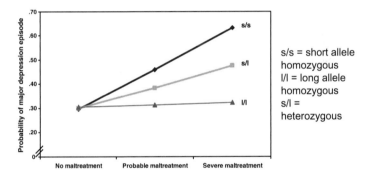

PLATE 3. Effect of maltreatment in childhood on liability to depression moderated by *5HTT* gene (from Caspi et al 2003). A full-color version of this figure is available in the color plate section of this book.

In each case, there was no significant main effect for G, a small but significant main effect for E, but a much larger effect for G × E. Four key background features were involved in the choice of strategy. First, the phenotype to be studied was selected as one in which quantitative genetic evidence suggested the likelihood of G × E. Second, the gene studied was one for which there was some, albeit inconsistent, evidence of genetic effects on that phenotype. Third, the evidence on the biology of the phenotype, and on the biology of the genetic effects suggested a plausible pathway for G × E. Fourth, there was good evidence of environmentally mediated causal effects from the E to be studied. Various genetically informed natural experiments provided the means to test such a causal inference (Rutter 2007b).

Moffitt and Caspi (see references above) have argued strongly for a hypothesis-testing approach and for a strategy starting with identified specific risk environments. The logic of the first point is that gigantic 'fishing expeditions' carry the huge risk of many false positive findings. Moreover, if the interest is, as it has to be, on the biological meaning of G × E, it makes sense to start from a base that involves some biological understanding of possible pathways.

The second point runs counter to the usual genetic assumption that it is only genes that are involved in the basic biology and that, therefore, they should provide the starting point. Once there is sufficient molecular evidence on the genes that influence sensitivity to the environment, they may constitute the best starting point, but we lack that evidence at the moment. Hence, the recommendation that, at the present time, the preferable starting point should be environments with demonstrated environmentally mediated causal effects. It should be added, of course, that the assumption that it is only genes that affect the basic biology is wrong (see, for example, the effects of institutional deprivation on brain growth [Rutter et al 2007]; the effects of environments on biological programming [Rutter 2006c]; and the effects of environments on gene expression [Bird 2007, Meaney & Szyf 2005, Mill & Petronis 2007].

Testing for the internal validity of G × E within a single study

With any claim to have identified G × E, the first need is to examine the internal validity of the claim; namely, the evidence for G × E in the sample studied, with the phenotype being considered, and with the specific identified and measured G and E. Four crucial issues have to be considered. First, could the interaction be an artefact deriving from the scaling approach used? Some critics have seemed to suggest that just because interactions can be artefactual, it means that artefact should be assumed (e.g. Eaves 2006). We need to reject that destructive counsel of despair. How can the scaling problem be addressed? The Dunedin group took three steps. Let me use the 5HTT (serotonin transporter) depression finding as an example. First, they used a range of different measures of the phenotype and found that the G × E applied across that range. Similarly, they checked whether the applicable E concerned only contemporaneous negative life events; findings showed that it did not. It included much earlier physical maltreatment. Second, they replaced their initial categorical measure with a dimensional equivalent and again found the same G × E. Third, they selected a polymorphism with scaling properties that were closely similar to the polymorphism being studied, but *without* the same relevant biological effects. Importantly, they found that this G showed no G × E. We need to consider whether these tests of possible scaling artefact were adequate and if comparable steps should be required for other G × E claims.

The second issue concerns the confidence that the E being studied actually involves an environmentally mediated causal effect. That query applies across the whole of medicine and it is clear that a series of crucial methodological issues need to be tackled before there can be confidence in the causal inference (see Academy of Medical Sciences 2007, Rutter 2007b, for a review of these). Ordinarily, the causal inference will need to be investigated *before* embarking on the G × E investigation. Once again, we will need to consider the extent to which this issue applies across G × E research and, if it applies widely, how it should be tackled.

The third query is whether the supposed G × E might actually reflect a gene–environment correlation (rGE). Separating rGE and G × E has been a major challenge in quantitative genetic research but the situation is rather different with molecular genetics. No longer is it necessary to infer rGE; rather, it can be examined directly with respect to the identified G and the identified E. Of course, the possibility should always be examined but, at least in the Dunedin studies, rGE has not been found in relation to the G and E investigated for G × E. More broadly, however, we need to ask whether it is sensible to expect rGE; after all, genes do not code for environments. Rather, they code for chemical effects that could influence behavior. It has been argued that we need to shift focus from rGE to genetic effects on behaviors that might play a role on the shaping and selecting of environments (Rutter 2006a). Such genetically influenced behaviors could mimic G × E but the process to be investigated must concern the *behaviors* that shape/select environments, and not environments as such.

The fourth challenge is to deal with the possibility that the interaction actually represents G × G, rather than G × E. In other words, the query is whether the E is representing a genetic liability (arising from rGE). This is less likely if there is no rGE with the allelic variation being studied, but there could still be G × G if the E was influenced by *other* genes. That possibility is difficult to exclude if the G × E has no observable time frame, but it is testable if it has. Thus, the Dunedin group argued that, whereas G × E (in the case of 5HTT, life events and depression) *had* to apply only to life events occurring *before* the onset of depression, G × G should operate equally regardless of the timing of life events. Their findings showed *no* G × E in the case of post-onset life events, making the possibility of G × G implausible. We will need to consider whether other tests are needed and whether similar testing has been undertaken with G × E in relation to other phenotypes.

Testing the external validity of G × E as found in other populations

The necessary next question, if the tests of internal validity have been passed, is whether there is *external* validity in the sense that the observed G × E represents a general effect that extends to other populations. The usual approach has been to ask if the finding can be replicated by other investigators using different samples. Sometimes, too, meta-analyses are undertaken to determine if there is still a significant effect when all studies are combined. Both have a place, but both have limitations. Thus, meta-analyses that pool samples rather than individuals are tremendously vulnerable to the vagaries of different measures and divergent samples. Replications share the same problems but with the additional query of what inferences to draw from occasional non-replications. So far as the two initial Dunedin studies are concerned, both approaches have been adopted. Thus, the MAOA finding was subjected to a meta-analysis that confirmed the significant G × E (Kim-Cohen et al 2006, Taylor & Kim-Cohen 2007), the statistical significance being evident with or without the original Dunedin study. The 5HTT finding has been examined in 20 studies to date, with 17 positive replications and just three non-replications. Of course, the studies vary in their rigor, their measures, and the nature of the samples studied. Nevertheless, the degree of replication is most impressive compared with the usual story in psychiatric genetics of numerous failures to replicate, with only the occasional positive confirmation. Should we adopt a football score approach such that 17 'for' and 3 'against' clearly means that the 'for's win? I suggest not (Rutter 1974). Of course, chance will operate so that the negatives could arise in that way. But, always, we need to adopt a more critical approach. In many circumstances, publication bias constitutes a major problem (i.e. positive findings are more likely to be published than negative findings—see Academy of Medical Sciences 2007; also Turner et al 2008). In this case, that is not a credible explanation because of the hostility of the behavioral genetic establishment, many of whose members are only too keen to rubbish the Dunedin findings.

Accordingly, it is necessary to approach the matter another way. Are the non-replications from reputable studies of adequate size? They seem to be. Are the confirmations solid? The main query in that connection stems from the fact that some of the confirmations are only partial; for example, they apply only to females, or they apply only to less severe life events. Should these be automatically discounted? Obviously not. Other research evidence clearly points to both age and gender differences in the genetic liability for depression. Also, of course, medical genetics has been consistent in showing that genetic heterogeneity across, as well as within, populations is usual. Should we, therefore, just assume that these variations account for the partial replications and the non-replications? I think not. Rather, we need to examine the totality of the evidence. In our discussions, we will need to consider how best to do that—not just with respect to the Dunedin findings but also in relation to the other phenotypes that we will be considering. Does study of other medical conditions provide guidance on how we should proceed?

Biological meaning of G × E

The final crucial issue concerns the biological meaning of G × E—the overriding goal of the G × E research. In a real sense, of course, if G × E can be shown to

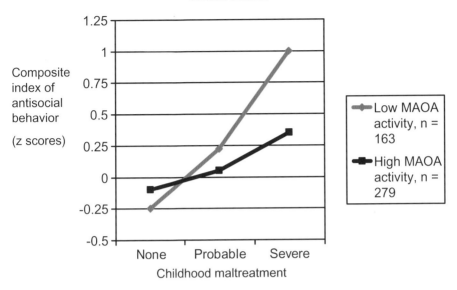

PLATE 1. Antisocial behavior as a function of MAOA activity and a childhood history of maltreatment (from Caspi et al 2002).

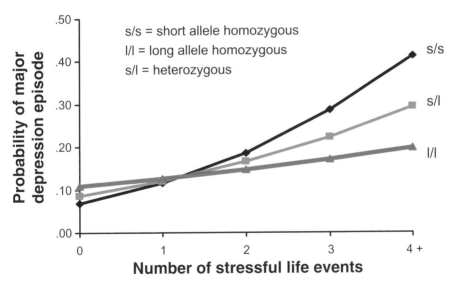

PLATE 2. Effect of life stress on depression moderated by *5HTT* gene (from Caspi et al 2003).

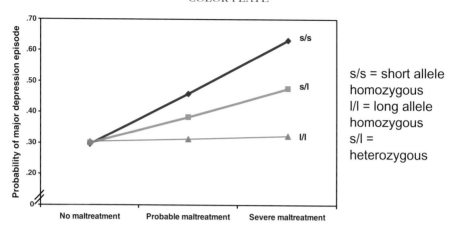

PLATE 3.   Effect of maltreatment in childhood on liability to depression moderated by *5HTT* gene (from Caspi et al 2003).

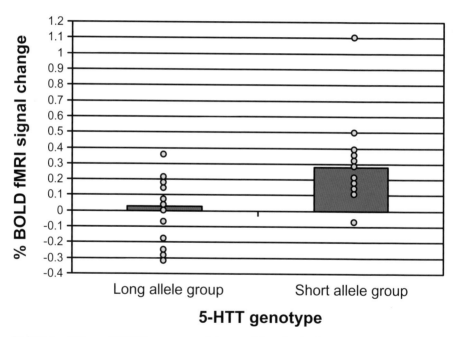

PLATE 4.   Effects of *5HTT* genotype on right amygdala activation in response to fearful stimuli (from Hariri et al 2002).

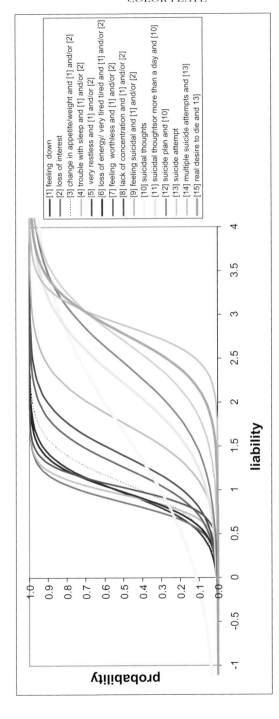

PLATE 5. Item response curves for each item using estimated values of *a* and *b* of equation [1]. The x-axis represents the normally distributed trait, liability to depression and the y-axis is the probability of endorsement of an item.

have biological correlates, this also provides an important source of external validation. So far as the Dunedin 5HTT finding is concerned, the biology has been studied in two main ways (Caspi & Moffitt 2006). First, studies of rhesus monkeys by Suomi's research group (Bennett et al 2002, Champoux et al 2002) showed the same G × E found in humans and showed cerebrospinal fluid (CSF) correlates. The evidence provides persuasive support for the Dunedin findings but three queries have to be considered. Is the gene in rhesus monkeys homologous with the human gene? Is peer rearing (which is known to be a major risk factor for adverse reactions in rhesus monkeys) comparable with the physical abuse studied in humans? To what extent is the phenotypic outcome in monkeys comparable with that in humans? It will be appreciated that these are the types of concerns that apply to any use of animal models. We will need, therefore, to consider how much reliance should be placed on such models in studying G × E with any of the medical conditions that we discuss in this symposium.

Second, the Weinberger/Hariri/Meyer-Lindenberg group at the National Institutes of Health has pioneered the use of structural brain imaging with experimental paradigms in humans to investigate G × E. The approach has three very important strengths: (1) rather than relying on reported E, it uses an experimental induction of the relevant E; (2) similarly, rather than relying on clinical disorder outcomes, it assesses effects on the brain; and (3) to avoid the possible problem that the G × E is somehow tapping a G effect on some medical condition, the samples were deliberately chosen to *exclude* individuals with any known psychopathology (Plate 4). The findings have been striking in showing strong neural correlates of G × E, using

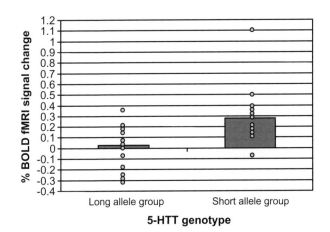

PLATE 4.   Effects of *5HTT* genotype on right amygdala activation in response to fearful stimuli (from Hariri et al 2002). A full-color version of this figure is available in the color plate section of this book.

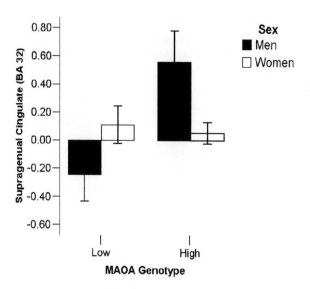

FIG. 2.   Genotype effect on anterior cingulate activation during response inhibition (data from Meyer-Lindenberg et al 2006).

samples of quite modest size (Hariri et al 2002). Moreover, the findings have since been confirmed in a later study by the same group (Hariri et al 2005) and replicated by others (Heinz et al 2005). Note, however, that as with the Dunedin study replications, the replication findings sometimes differ in details.

Sometimes, too, the differences are more substantial. For example, the Meyer-Lindenberg et al (2006) study of the MAOA (monoamine oxidase) G × E found major effects in males but no significant effects in females (Fig. 2). Once more, we have to ask ourselves what these differences might mean, and whether they raise queries about the reality of the biological effects. We will also need to recognize that these brain imaging findings, impressive though they are, still fall short of elucidating the biological mechanisms involved in G × E. We must consider what other kinds of research strategy will need to be employed.

These are just some of the questions that we will need to be debating over the next two days. It is not, of course, that anyone expects that our deliberations will lead to a nice, neat package of answers but, hopefully, they should provide useful guidance on the possible ways forward. I am very conscious that, in seeking to lay out our agenda, I have strayed in a major way into the papers still to come, but my intended purpose is to stimulate you to think about the questions that we will need to consider across all the papers. By the end of the symposium I hope that we may be in a stronger position to assess the current state of knowledge but, more importantly, we should strive to identify the ways forward.

References

Academy of Medical Sciences 2007 Identifying the environmental causes of disease: how should we decide what to believe and when to take action? London: Academy of Medical Sciences

Bennett AJ, Lesch KP, Heils A et al 2002 Early experience and serotonin transporter gene variation interact to influence primate CNS function. Mol Psychiatry 7:118–122

Bird A 2007 Perceptions of epigenetics. Nature 447:396–398

Brown GW, Harris TO 1978 The social origins of depression: a study of psychiatric disorder in women. London: Tavistock

Caspi A, Moffitt TE 2006 Gene-environment interactions in psychiatry: joining forces with neuroscience. Nat Rev Neurosci 7:583–590

Caspi A, McClay J, Moffitt TE et al 2002 Role of genotype in the cycle of violence in maltreated children. Science 297:851–854

Caspi A, Sugden K, Moffitt TE et al 2003 Influence of life stress on depression: moderation by a polymorphism in the 5-HTT gene. Science 301:386–389

Caspi A, Moffitt TE, Cannon M et al 2005 Moderation of the effect of adolescent-onset cannabis use on adult psychosis by a functional polymorphism in the COMT gene: longitudinal evidence of a gene-environment interaction. Biol Psychiatry 57:1117–1127

Caspi A, Williams B, Kim-Cohen J et al 2007 Moderation of breastfeeding effects on the IQ by genetic variation in fatty acid metabolism. Proc Natl Acad Sci USA 107:18860–18865

Champoux M, Bennett A, Shannon C, Higley JD, Lesch KP, Suomi SJ 2002 Serotonin transporter gene polymorphism, differential early rearing, and behavior in rhesus monkey neonates. Mol Psychiatry 7:1058–1063

Crabbe JC, Wahlsten D, Dudek BC 1999 Genetics of mouse behavior: interactions with laboratory environment. Science 284:1670–1672

Eaves LJ 2006 Genotype x environment interaction in psychopathology: fact or artifact? Twin Res Hum Genet 9:1–8

Greenland S, Rothman KJ 1998 Concepts of interaction. In: Winters R, O'Connor E (eds) Modern epidemiology, Second Edition. Lippincott-Raven: Philadelphia, p 329–342

Hariri AR, Mattay VS, Tessitore A et al 2002 Serotonin transporter genetic variation and the response of the human amygdala. Science 297:400–403

Hariri AR, Drabant EM, Munoz KE et al 2005 A susceptibility gene for affective disorders and the response of the human amygdala. Arch Gen Psychiatry 62:146–152

Heinz A, Braus DF, Smolka MN et al 2005 Amygdala-prefrontal coupling depends on a genetic variation of the serotonin transporter. Nat Neurosci 8:20–21

Kim-Cohen J, Caspi A, Taylor A et al 2006 MAOA, maltreatment, and gene–environment interaction predicting children's mental health: new evidence and a meta-analysis. Mol Psychiatry 11:903–913

Mackay TFC, Anholt RRH 2007 Ain't misbehavin'? Genotype-environment interactions and the genetics of behavior. Trends Genet 23:311–314

Meaney MJ, Szyf M 2005 Environmental programming of stress responses through DNA methylation: life at the interface between a dynamic environment and a fixed genome. Dialogues Clin Neurosci 7:103–123

Meyer-Lindenberg A, Buckholtz JW et al 2006 Neural mechanisms of genetic risk for impulsivity and violence in humans. Proc Natl Acad Sci USA 103:6269–6274

Mill J, Petronis A 2007 Molecular studies of major depressive disorder: the epigenetic perspective. Mol Psychiatry 12:799–814

Moffitt TE, Caspi A, Rutter M 2005 Strategy for investigating interactions between measured genes and measured environments. Arch Gen Psychiatry 62:473–481

Moffitt TE, Caspi A, Rutter M 2006 Measured gene-environment interactions in psychopathology: concepts, research strategies, and implications for research, intervention, and public understanding of genetics. Perspect Psychol Sci 1:5–27

Plomin R, DeFries JC, Fulker DW 1988 Nature and nurture during infancy and early childhood. New York: Cambridge University Press

Rutter M 1974 Epidemiological strategies and psychiatric concepts in research on the vulnerable child. In: Anthony E, Koupernik C (eds) The child and his family: children at psychiatric risk (Vol. 3). New York: Wiley, p 167–179

Rutter M 1983 Statistical and personal interactions: facets and perspectives. In: Magnusson D, Allen V (eds) Human development: an interactional perspective. New York: Academic Press, p 295–319

Rutter M 2006a Genes and behavior: nature–nurture interplay explained. Oxford: Blackwell

Rutter M 2006b Implications of resilience concepts for scientific understanding. Ann N Y Acad Sci 1094:1–12

Rutter M 2006c The psychological effects of early institutional rearing. In: Marshall P, Fox N (eds) The development of social engagement: neurobiological perspectives. New York & Oxford: Oxford University Press, p 355–391

Rutter M 2007a Gene-environment interdependence. Dev Sci 10:12–18

Rutter M 2007b Proceeding from observed correlation to causal inference: the use of natural experiments. Perspect Psychol Sci 2:377–395

Rutter M, Pickles A 1991 Person–environment interactions; concepts, mechanisms, and implications for data analysis. In: Wachs TD, Plomin R (eds) Conceptualization and measurement of organism–environment interaction. Washington, DC: American Psychological Association, p 105–141

Rutter M, Moffitt TE, Caspi A 2006 Gene–environment interplay and psychopathology: multiple varieties but real effects. J Child Psychol Psychiatry 47:226–61

Rutter M, Beckett C, Castle J et al 2007 Effects of profound early institutional deprivation: an overview of findings from a UK longitudinal study of Romanian adoptees. Eur J Dev Psychol 4:332–350

Tabery J 2007 Biometric and developmental gene–environment interactions: looking back, moving forward. Dev Psychopathol 19:961–976

Taylor A, Kim-Cohen J 2007 Meta-analysis of gene-environment interactions in developmental psychopathology. Dev Psychopathol 19:1029–1037

Turner EH, Matthews AM, Linardatos E, Tell RA, Rosenthal R 2008 Selective publication of antidepressant trials and its inference on apparent efficacy. N Engl J Med 358:252–260

Wachs TD, Plomin R 1991 Conceptualization and measurement of organism–environment interaction. Washington, DC: American Psychological Association

Wahlsten D, Bachmanov A, Finn DA, Crabbe JC 2006 Stability of inbred mouse strain differences in behavior and brain size between laboratories and across decades. Proc Natl Acad Sci USA 103:16364–16369

2. Gene–environment interaction: overcoming methodological challenges

Rudolf Uher

MRC Social, Genetic and Developmental Psychiatry Research Centre, Institute of Psychiatry, King's College London, UK

Abstract. While interacting biological effects of genes and environmental exposures (G × E) form a natural part of the causal framework underlying disorders of human health, the detection of G × E relies on inference from statistical interactions observed at population level. The validity of such inference has been questioned because the presence or absence of statistical interaction depends on measurement scale and statistical model. Furthermore, the feasibility of G × E research is threatened by the fact that tests of statistical interaction require large samples and their power is substantially reduced by unreliability in the assessments of genes, environmental exposures and pathology. It is demonstrated that concerns about statistical models and scaling can be addressed by integration of observational and experimental data. Judicious selection of genes and environmental factors should limit multiple testing. To overcome the challenge of low statistical power, it is suggested to maximize the reliability of measurement, integrate prior knowledge under Bayesian framework and facilitate pooling of data across studies by use of standardized stratified reporting. Consistencies and discrepancies among studies can be exploited for methodological analysis and model specification.

2008 Genetic effects on environmental vulnerability to disease. Wiley, Chichester (Novartis Foundation Symposium) p 13–30

The motivation to study gene–environment (G × E) interactions

The integration of genetic and environmental investigations that occurred over the last decade was motivated by the disappointing results of the search for genetic causes and discrepancies between findings of genetic and environmental epidemiology. The lack of replicable gene–disease associations for most common diseases (Ioannidis et al 2001) contrasted with twin studies that indicated large genetic contributions. Most twin studies allocated little importance to shared environment, contradicting the epidemiological evidence for large effects of environmental factors shared by family members (e.g. social class). A possible explanation is that in a twin study a G × E involving shared environment is attributed to genes, but in an association study, detection of genetic effect depends on variability in

13

environmental exposure. Characterization of genes and environment in the same sample has the potential to explain inconsistencies and improve replicability. For example, there was a strong rationale for the length polymorphism in the serotonin transporter gene (5HTTLPR) as a candidate risk factor for depression, but results of association studies were inconsistent. It was found that 5HTTLPR moderates the depressogenic effect of environmental factors, including stressful life events and child abuse (Caspi et al 2003). Unlike previous association studies, this G × E has a favorable replication record (Uher & McGuffin 2007).

While many authors reserve the term 'gene–environment interaction' for the relatively few identified instances of statistically significant interactions between genes and measured exposures, I propose that biological G × E is a universal mechanism in the genesis of health and disease. One reason for this proposition is that genetic causes (indicated by twin and adoption studies) and environmental causes (indicated by epidemiological studies and secular trends) appear to add up to explain much more than 100% of variance in outcomes. This super-additivity implies biological interaction where genes and environment are component causes, each of them necessary but not sufficient to produce a disorder (Rothman & Greenland 2005). The other reason is the conceptual consideration that varying rates of environmental exposure (E) can transform a statistical interaction into main effect of genes (G) or no effect (Fig. 1). If the effects of G and E are crossed, i.e. genotype that confers susceptibility to an environmental pathogen confers an advantage in the absence of exposure, the same biological mechanism may account for opposite findings of gene–disease associations in populations with very high and very low rates of exposure. In conjunction with the low power of statistical tests to detect interactions, this suggests that false negative findings in G × E research are more common than either true or false positives.

While it may be obvious that study of G × E is worthwhile, it is also apparent that research of the evasive G × E presents some formidable challenges. In the next paragraphs I will provide an overview of methodological requirements for G × E studies. Some of them will be discussed in detail in other contributions to the symposium.

Candidate selection

Multiple testing is a well known problem of genetic association studies and is, literally, multiplied in the G × E research. The probability of false positives at a specified level of probability increases with the number of tests and if factors with low prior probability of true effect are included, false positives may be more likely than true positives (Ioannidis et al 2001). Therefore, researchers must perform judicious selection of candidate genes and exposures. The guiding principle is biological plausibility. However, judgment on biological plausibility has a subjective compo-

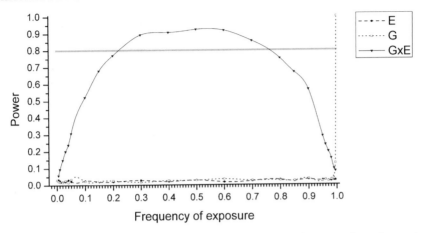

FIG. 1. Influence of variation in exposure frequency on statistical power to detect interaction and main effects of genes and environment. Each point is based on 10000 simulations of samples of 1000 drawn from a population with equal distributions of two genotypes with a G × E interaction of moderate effect size (0.3) and no main effects. Grey line corresponds to the power of 0.8. In samples with frequency of exposure close to 0, there is no detectable interaction or main effect. For exposure frequency between 0.24 and 0.76, a sample of 1000 provides sufficient power to detect interaction. If only exposed individuals are sampled (exposure frequency = 1), the interaction transforms into main effect of genotype. The probability of detecting a spurious main effect of environment remains at the 2.5% chance level across the range of exposure frequency as long as the interaction term is retained in the equation.

nent and therefore it is helpful to follow a systematic approach (Hunter 2005). There are several approaches to the selection of candidates.

The first approach is to select a polymorphic gene with a known main effect and an environmental factor with evidence indicating a biological interaction with that polymorphism. An example is the APOE-ε4 allele of the apolipoprotein E gene (a known risk factor for cognitive decline) and estrogen use (biological interaction between estrogen and APOE was demonstrated *in vitro*) in a study on cognitive decline in elderly women. An interaction was demonstrated with a protective effect of estrogens restricted to women without the at-risk genotype (Yaffe et al 2000).

A second approach is to select an environmental pathogen with a known main effect and a polymorphism in a gene with a rationale to indicate biological interaction. This approach has led to the discovery of interaction between stressful life events and 5HTTLPR in the etiology of depression (Caspi et al 2003). While stressful life events are a recognized trigger of depression, their effect depends on individual's vulnerability (Brown & Harris 1978). The 5HTTLPR polymorphism has been shown to affect the production of serotonin transporter and to interact with experimental stress in non-human primates.

A third approach is to choose genetic and environmental factors that are both known to increase the risk of a disease. An example is mutation in the gene for the coagulation factor V and oral contraceptive use in deep venous thrombosis. While presence of one of the factors increased the risk four or eight times, their co-occurrence increased the risk of thrombosis 35-fold (Vanderbroucke et al 1994). When there are main effects of both genotype and exposure, the presence or absence of statistical interaction depends on measurement scale (Rutter 2008, Coventry et al 2008). The findings of Vanderbroucke et al (1994) are consistent with interaction when an additive model is assumed but can also be described as multiplicative effects with no interaction term (Clayton & McKeigue 2001). However, scale dependence of statistical interaction does not diminish the impact of the finding, e.g. for counseling on the use of oral contraception among carriers of the coagulation factor mutation.

The three approaches described and exemplified above have proven successful. However, it is possible that crossed interactions exist where neither the genetic nor the environmental factor has consistent main effect. Clues to such interactions may be provided by variable main effects across studies and *in vitro* or animal model evidence of biological interaction. However, search for 'pure' G × E without main effects will remain a difficult territory.

Study design

Several types of studies allow for testing of G × E, each having its own advantages and problems (Andrieu & Goldstein 1998; Table 1). Cohort study is the design of choice for common diseases. The assessment of environmental exposure occurs before the onset of disease and therefore is relatively free of information and selection bias. The disadvantage of cohort studies is the need to follow up a large number of individuals over a long period. Less common diseases and diseases with late or widely distributed age of onset would require impractically large cohort samples. Incomplete follow-up may be a problem, especially if loss to follow-up is associated with the exposure (e.g. social class, employment) or the outcome (e.g. mental illness).

Case-control studies are the economical alternative to cohorts and are the design of choice for less common outcomes. Methods of case ascertainment and case-control matching are essential to minimize selection bias. If cases are ascertained from a clinical setting, then the researchers may inadvertently be exploring determinants of treatment-seeking behavior rather than disease itself. This is a major confounder especially in less severe disorders where a majority of cases may not seek medical care. In case-control studies, the exposure is usually assessed retrospectively. Therefore, these studies are also prone to information bias. Information bias can be minimized by seeking an objective record of

TABLE 1   Designs for gene–environment interaction studies

Design	Advantages	Disadvantages
Cohort	Robust to selection bias Robust to information bias	Cost Duration Incomplete follow-up Low power for rare outcomes Genetic contribution to exposure Population stratification
Case-control	Economic Powerful Suitable for rare outcomes	Selection bias Information bias Genetic contribution to exposure Population stratification
Cases only	No controls required	Selection bias Information bias Does not test main effects Untestable assumption of no rGE Genetic contribution to exposure Population stratification
Family	Robust to population stratification	Cost Low power Selection bias Information bias Genetic contribution to exposure

Experimental	Exposure manipulated
	  independently of genes
		Limited to benign exposures and
		  animal models
		Selection bias
		Population stratification

exposure or obtaining information on exposure and outcome from different sources.

On the assumption that environmental exposure is independent of genotype in the population, G × E can be tested in cases only as presence of simple association between genotype and exposure (Khoury & Flanders 1996). However, correlation between genotype and environment appears to be the rule rather than exception and cases-only design does not allow testing this crucial assumption. The value of cases-only design is further reduces by its failure to measure main effects of genotype and environment. As G × E can be transformed into a main effect of genes (Fig. 1), this is a major shortcoming. Therefore, while cases-only design may be attractive for researchers working in clinical setting, it fails to provide a comprehensive test of causal mechanism.

Family-based designs eliminate the potential for confounding by population stratification (i.e. recent mixture of populations resulting in subgroups that differ on both genotype and outcome). However, with the advent of effective tools to control for population stratification (Devlin et al 2001) and evidence that stratification may not play major role in the majority of instances (Wellcome Trust Consortium 2007), family-based designs are becoming less attractive. Both genes and large part of environment are shared between family members. This gene–environment correlation by design renders family studies less powerful for detection of interaction between genetic and environmental factors (Teng & Risch 1999). Twin samples belong to family-based designs and share their advantages and shortcomings.

The designs reviewed above are observational and cannot control for gene–environment correlation (rGE). Experimental designs where environmental factors can be manipulated independently of genes are crucial to overcome this limitation of the epidemiological method. Experimental designs are used in animal studies, analogue human studies and in pharmacogenetics. These will be covered in the paragraph on biological validation.

Power and sample size

Any study that purports to 'test' a G × E must have adequate statistical power to do so. But the issue of power in G × E studies is a complex one. Power depends on sample size, strength of the interaction, variability in environmental exposure, distribution of genotypes in population, and the reliability of measurement of exposure, genotype and outcome (Luan et al 2001). As the effect size usually appears larger in the first study reporting an effect, power calculation based on an initial study may be over-optimistic (Ioannidis et al 2001). All too often, power is equated with sample size and the other factors are neglected. The importance of exposure distribution, genotype frequency and measurement accuracy is demonstrated in simulation studies. Figure 1 shows how variations in frequency of environmental exposure impact on power: if between 24% and 76% of subjects are exposed, a sample of 1000 with equal distributions of genotypes provides adequate power to detect a moderately strong interaction. Figure 2 shows the effect of genotype frequency on power to detect interaction of moderate effect size (0.3). While small inequality in genotype frequencies has only moderate influence, interactions involving genotypes with frequencies of less than 10% of population are extremely difficult to detect. For example, interaction of moderate effect size involving a genotype that is present in only 5% of the population would require a sample of 5200 to achieve a power of 0.8. This means that interactions involving less common genetic variants are only detectable if they have strong effect. Figure 3 demonstrates the influence of measurement accuracy on statistical power to detect an interaction of moderate effect size involving a genotype present in 20% of the population. Decrease of 0.2 in the reliability of environment and outcome measures equates to loosing nearly halve the sample. This can easily

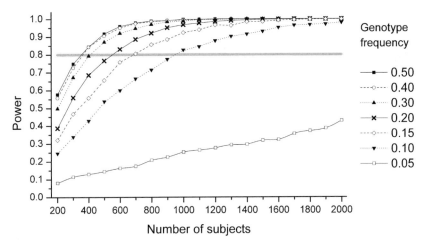

FIG. 2.   The influence of genotype frequency on the relationship between power and sample size. Each point is based on 10 000 simulations of samples with a moderately strong G × E (difference of correlations 0.3) and 100% reliability of measurement. Grey line corresponds to the power of 0.8.

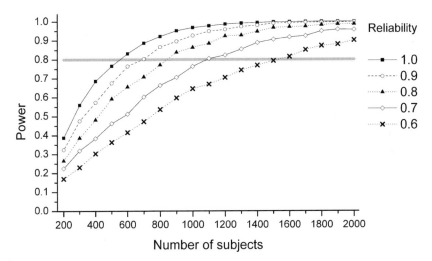

FIG. 3.   The influence of measurement reliability on the relationship between power and sample size. Each point is based on 10 000 simulations of samples with a moderately strong G × E (difference of correlations 0.3) and minor genotype frequency of 20%. Grey line corresponds to the power of 0.8.

occur, e.g. when short self-report questionnaires are used instead of detailed interviews. Consequently, smaller studies with accurate measurement may be preferable to large studies with less reliable measures (Wong et al 2003). These issues need to be considered when planning a study and balancing sample size and unit cost.

Measurement of genes, exposures and pathology

The simulation results in Fig. 2 highlight the importance of measurement reliability. This concerns measurement of genotype, environmental exposure and the outcome. Measurement of genotype is prone to genotyping error, which leads to loss of power and, due to its non-random character, to confounding. Methods for quantifying and controlling genotyping error have been described (Pompanon et al 2005). Loss of power also occurs if a measured polymorphism is a proxy for the true causative genetic factor, e.g. via a linkage disequilibrium. For example, opposite results of pharmacogenetic studies of antidepressants in European and Asian populations may be pointing to the fact that the measured polymorphism is not the true causative factor (Serretti et al 2007). Examination of all polymorphisms in the region of interest and their combinations in haplotypes should be considered in exploring a G × E. The complex issue of combining multiple genetic markers and megavariate data in the context of whole-genome association is examined in a separate contribution (Snieder et al 2008).

Environmental exposures such as social adversity, air pollution or dietary intake are quantitative: there may be more or less of them and they can be measured on a continuous scale. Other exposures occur in discrete doses (e.g. oral contraception) or deviate so much from the ordinary that they can be considered as yes-or-no events (e.g. a natural catastrophe). Similar considerations apply to outcomes: obesity and high blood pressure are quantitative deviations from normality, but stroke, deep vein thrombosis and acute leukemia occur as discrete events. Some conditions, such as depression, can be conceptualized as either dimensional or categorical and relative merits of the two conceptualizations are subject of debates (Kraemer et al 2004). Dichotomous outcomes may appear attractive as they conform to the biomedical model and allow simple visual presentation of results. However, dichotomization of naturally continuous variables leads to a loss of information and power that is commensurate with throwing away one third of the sample (Royston et al 2006). Therefore, for purposes of statistical testing, it is almost always preferable to use continuous measures. Exceptions may include instances where clear bimodality is present in the data or there is strong prior evidence of a 'tipping-point' in the relationship between exposure and outcome. Such relationship has been proposed for stressful life events and depression where only events with severe contextual threat appear to be related to depression onset and additivity between events is limited (Brown & Harris 1978).

The G × E concept is based on the assumption that environment and genes are independent. While this assumption appears reasonable, it is often violated. For example, stressful life events have been conceptualized as environmental factors, but there is a significant genetic contribution to reported events (Kendler & Baker 2007). Some authors went so far as using the personality trait neuroticism as a proxy for genetic vulnerability (Kendler et al 2003) or environmental adversity (Jacobs et al 2006). As neuroticism is determined by both genes and environment, this substitution clouds the etiological exploration. A more constructive approach is the use of refined measures of environment: an interview that quantifies the level of independence of stressful life events (Brown & Harris 1978) or objectively recorded events in natural experiments (Kilpatrick et al 2007).

Factors that are considered as environmental, e.g. smoking, are strongly determined by personality and genetic factors. Personality-related factors and stressful life events also influence detection of physical health outcomes including abdominal pain, appendectomy, peptic ulcer or diabetes control (Creed 2000). Therefore, G × E research on physical health may benefit from taking the same precautions to take account of confounders related to selecting environments, recalling and reporting exposures and outcomes.

Statistical inference

The distinction between biological and statistical interaction has been highlighted in the introduction chapter (Rutter 2008). The biological G × E is the causal mechanism of interest, but its detection in human studies relies on inference from statistical interactions. The validity of such inference has been questioned as the presence or absence of statistical interaction depends on measurement scale and statistical model (Clayton & McKeigue 2001). Consequently, the debate over G × E often focuses on statistical issues and overlooks the fact that it is the biological causative mechanism that is of primary interest. I would like to propose that G × E research should rely less on statistical inference from single studies and more on experimental validation of elements of the biological causative pathway and synthesis of findings across studies on diverse populations.

Statistical tests for interactions of continuous and categorical variables have been described (Aiken & West 1991). The aspects of analysis specifically relevant to the G × E are testing for rGE and controlling for population stratification (Devlin et al 2001). These two issues should always be addressed.

As the theory of disease causation converges on a consensus that a simple additive model corresponds to biological independence, it is likely that previous emphasis on a multiplicative model has led to the omission of existing interaction (Rothman & Greenland 2005, Rutter 2008). In earlier paragraphs, I have demonstrated that lack of statistically significant interaction can have numerous explanations other than absence of biological interaction, including unequal distributions of genotypes and exposures and insufficient reliability of measurement. While spuriously positive findings of G × E can occur (Eaves 2006), the generally low power of statistical models to detect interaction indicates that false negative findings are much more common than false positive ones (Luan et al 2001).

Biological validation

Epidemiological studies are useful to indicate the presence of G × E and to quantify its contribution to illness. However, the epidemiological method is limited by its observational character and inability to separate genetic and environmental factors. Experimental methodology is required to manipulate environmental factors independently of genotype and to probe the elements of the hypothetical pathogenic pathway (Caspi & Moffitt 2006).

While it would be unethical to allocate human individuals to environmental adversity, experimental method can be applied to exposures with expected beneficial effects. This is done in randomized clinical trials that allow unbiased test of interaction between genotype and exposure to medication (Serretti et al 2007). Randomization to non-pharmacological intervention remains to be exploited in G × E research. Adverse environmental factors can be experimentally manipulated in animal models. A complementary approach is exploration of endophenotypes in human volunteers including reactions to harmless analogues of environmental stressors, e.g. brain reactivity to standard emotional stimuli as a proxy of vulnerability to adverse environment (Canli & Lesch 2007).

Caspi & Moffitt (2006) suggested that an iterative process between neuroscience and epidemiology will build a nomological framework for testing an etiological mechanism from multiple perspectives. The 5HTTLPR × environmental adversity interaction in depression exemplifies how synergistic evidence from human epidemiology, animal studies and endophenotype research provides support for G × E (Uher & McGuffin 2007).

Data synthesis

Most studies lack power to detect G × E of moderate effect. Large samples are sometimes collected with economical but less reliable methodology and as a result may not be more powerful than medium-size studies (Wong et al 2003).

It may therefore be impractical to expect each study to provide an independent test of a G × E. Integration of data across samples may provide insights that are not obtainable from a single study. A theoretical example is study A showing a main effect of gene in a population with ubiquitous exposure to environmental factor E, study B showing a G × E in a population with variable exposure to E and study C showing no effect in a sample where exposure to E is rare or buffered by a protective factor. Detailed analysis of such three studies would provide rich evidence of G × E in spite of superficially discordant results. This theoretical example also demonstrates the danger of overreliance on metaanalytic methods (a crude metaanalysis of these three studies would show no G × E).

In a methodological analysis of 18 studies that tested the interaction between 5HTTLPR and environmental adversity in the genesis of depression, we found methodological differences that have the potential to explain inconsistent results (Uher & McGuffin 2007). Sample age and gender composition and unreliable methods of measurement had the potential to explain negative findings in two large studies. The level of methodological differences was such that integration of results from these studies in a metaanalysis could have been misleading. Meta-analytic methods for groups of G × E studies that are sufficiently homogenous has been described (Taylor & Kim-Cohen 2007).

It is crucial that studies are reported in a way that facilitates systematic review and metaanalysis. Our review was complicated by inconsistent and incomplete reporting of results. Different authors assumed dominant or recessive models and reported the results on combined genotype groups that differed between studies. Results reported in such way are unsuitable for synthesis. To enable integration of knowledge across studies, reporting guidelines should be established (Ioannidis et al 2006). As a minimum requirement, descriptive statistics and correlation of outcome with exposure should be reported for each genotype group with heterozygotes separate from both homozygous groups.

When a new study aims to explore a previously identified G × E, there is a prior expectation what the results should be like. If this study is well designed and sufficiently powered, it can be considered as an independent replication. However, power may be less than expected and results may fall just above or just below the arbitrary threshold of statistical significance. It would be unwise to consider two very similar patterns of results as evidence for or against the G × E based on falling below or above the significance threshold. Bayesian approach provides an alternative framework where prior expectations (based on previous knowledge) can be incorporated into the statistical model and the new study is used to update the concept with additional data rather than provide an independent test. The departure of expectations in either direction is quantified as the change from prior to posterior probability.

Directions

In the near future, G × E research will move beyond the triangular single-gene to single-exposure to single-disorder relationship. An example of replicable three-way interaction is the G × G × E between 5HTTLPR, brain-derived neurotrophic factor (BDNF) and social adversity interaction (Kaufman et al 2006, Kim et al 2007, Wichers et al 2007). The genome of gut microbes may also interact with the genome of their human carrier and environmental factors, as has been demonstrated for paracetamol toxicity (Clayton et al 2006). However, a systematic extension of the G × E methodology to involve more than the familiar triangle will be a challenge for the years to come. I suggest that a balance between epidemiology, statistics, clinical and laboratory science will be essential.

The G × E will influence the concept of disease in etiological and clinical research. The case of a crossed G × E without main effects is especially intriguing as it implies etiological heterogeneity. For example, a proportion of depression cases are explained by an interaction between short alleles of the 5HTTLPR and environmental adversity but the prevalence of depression is not increased among carriers of the short alleles (not even with increasing age when exposure to severe life events accumulates). Among carriers of the long alleles, depression is just as common, but its causation is less understood. This crossed G × E should instigate a search for complementary causative mechanisms that explain depression among long-allele homozygotes. It is also possible that such heterogeneity of causative mechanisms will have implications for treatment. In the case of depression, certain individuals respond much better to psychological treatment and other respond better to medication. Will a measure of G × E contribution to aetiology predict differential response to these treatments? As the G × E research advances, I expect similar questions being raised in other medical disciplines.

References

Aiken LS, West S 1991 Multiple regression: testing and interpreting interactions. Sage Publications, Thousand Oaks

Andrieu N, Goldstein AM 1998 Epidemiologic and genetic approaches in the study of gene–environment interaction: an overview of available methods. Epidemiol Rev 20:137–147

Brown GW, Harris TO 1978 Social origins of depression. A study of psychiatric disorder in women. Routledge, London

Canli T, Lesch KP 2007 Long story short: the serotonin transporter in emotion regulation and social cognition. Nat Neurosci 10:1103–1109

Caspi A, Moffitt TE 2006 Gene–environment interactions in psychiatry: joining forces with neuroscience. Nat Rev Neurosci 7:583–590

Caspi A, Sugden K, Moffitt TE et al 2003 Influence of life stress on depression: moderation by a polymorphism in the 5-HTT gene. Science 301:386–389

Clayton D, McKeigue PM 2001 Epidemiological methods for studying genes and environmental factors in complex diseases. Lancet 358:1356–1360

Clayton TA, Lindon JC, Cloarec O et al 2006 Pharmaco-metabonomic phenotyping and personalized drug treatment. Nature 440:1073–1077

Creed F 2000 The study of life events has clarified the concept of psychosomatic disorders. In: Harris T (ed) Where inner and outer worlds meet. Routledge, London & New York p 275–287

Devlin B, Roeder K, Wasserman L 2001 Genomic control, a new approach to genetic-based association studies. Theor Popul Biol 60:155–166

Eaves LJ 2006 Genotype x environment interaction in psychopathology: fact or artifact? Twin Res Hum Genet 9:1–8

Hunter DJ 2005 Gene–environment interactions in human diseases. Nat Rev Genet 6:287–298

Ioannidis JP, Ntzani EE, Trikalinos TA, Contopoulos-Ioannidis DG 2001 Replication validity of genetic association studies. Nat Genet 29:306–309

Ioannidis JP, Gwinn M, Little J et al 2006 A road map for efficient and reliable human genome epidemiology. Nat Genet 38:3–5

Jacobs N, Kenis G, Peeters F et al 2006 Stress-related negative affectivity and genetically altered serotonin transporter function: evidence of synergism in shaping risk of depression. Arch Gen Psychiatry 63:989–996

Kaufman J, Yang BZ, Douglas-Palumberi H et al 2006 Brain-derived neurotrophic factor-5-HTTLPR gene interactions and environmental modifiers of depression in children. Biol Psychiatry 59:673–680

Kendler KS, Baker JH 2007 Genetic influences on measures of the environment: a systematic review. Psychol Med 37:615–626

Kendler KS, Gardner CO, Prescott CA 2003 Personality and the experience of environmental adversity. Psychol Med 33:1193–1202

Khoury MJ, Flanders WD 1996 Nontraditional epidemiologic approaches in the analysis of gene–environment interaction: case-control studies with no controls! Am J Epidemiol 144:207–213

Kilpatrick DG, Koenen KC, Ruggiero KJ et al 2007 The serotonin transporter genotype and social support and moderation of posttraumatic stress disorder and depression in hurricane-exposed adults. Am J Psychiatry 164:1693–1699

Kim JM, Stewart R, Kim SW et al 2007 Interactions between life stressors and susceptibility genes (5-HTTLPR and BDNF) on depression in Korean elders. Biol Psychiatry 62: 423–428

Kraemer HC, Noda A, O'Hara R 2004 Categorical versus dimensional approaches to diagnosis: methodological challenges. J Psychiatr Res 38:17–25

Luan JA, Wong MY, Day NE, Wareham NJ 2001 Sample size determination for studies of gene–environment interaction. Int J Epidemiol 30:1035–1040

Pompanon F, Bonin A, Bellemain E, Taberlet P 2005 Genotyping errors: causes, consequences and solutions. Nat Rev Genet 6:487–459

Rothman KJ, Greenland S 2005 Causation and causal inference in epidemiology. Am J Public Health 95 Suppl 1:S144–S150

Royston P, Altman DG, Sauerbrei W 2006 Dichotomizing continuous predictors in multiple regression: a bad idea. Stat Med 25:127–141

Rutter M 2008 Whither gene–environment interactions? In: Genetic effects on environmental vulnerability to disease. Wiley, Chichester (Novartis Found Symp) p 1–12

Serretti A, Kato M, De RD, Kinoshita T 2007 Meta-analysis of serotonin transporter gene promoter polymorphism (5-HTTLPR) association with selective serotonin reuptake inhibitor efficacy in depressed patients. Mol Psychiatry 12:247–257

Snieder H, Wang X, Lagou V, Penninx BWJH, Riese H, Hartman CA 2008 Role of gene–stress interactions in gene finding studies. In: Genetic effects on environmental vulnerability to disease. Wiley, Chichester (Novartis Found Symp) p 71–82

Taylor A, Kim-Cohen J 2007 Meta-analysis of gene–environment interactions in developmental psychopathology. Dev Psychopathol 19:1029–1037

Teng J, Risch N 1999 The relative power of family-based and case-control designs for linkage disequilibrium studies of complex human diseases. II. Individual genotyping. Genome Res 9:234–241

Uher R, McGuffin P 2008 The moderation by the serotonin transporter gene of environmental adversity in the aetiology of mental illness: review and methodological analysis. Mol Psychiatry 13:131–146

Vanderbroucke JP, Koster T, Briet E et al 1994 Increased risk of venous thrombosis in oral-contraceptive users who are carriers of factor V Leiden mutation. Lancet 344:1453–1457

Wellcome Trust Consortium 2007 Genome-wide association study of 14,000 cases of seven common diseases and 3,000 shared controls. Nature 447:661–678

Wichers M, Kenis G, Jacobs N et al 2008 The BDNF Val(66)Met x 5-HTTLPR x child adversity interaction and depressive symptoms: an attempt at replication. Am J Med Genet B Neuropsychiatr Genet 14:120–123

Wong MY, Day NE, Luan JA et al 2003 The detection of gene–environment interaction for continuous traits: should we deal with measurement error by bigger studies or better measurement? Int J Epidemiol 32:51–57

Wray N, Coventry W, James M, Montgomery GW, Eaves LJ, Martin NG 2008 Use of monozygotic twins to investigate the relationship between 5HTTLPR genotype , depression and stressful life events: an application of Item Response Theory. In: Genetic effects on environmental vulnerability to disease. Wiley, Chichester (Novartis Found Symp) p 48–67

Yaffe K, Haan M, Byers A et al 2000 Estrogen use, APOE, and cognitive decline: evidence of gene–environment interaction. Neurology 54:1949–1954

DISCUSSION

Heath: Let me speak in favor of also retaining a statistical concept of interaction as a way of being sensitive to our data. This will help us decide what we can infer from our data, and what we may be misleading ourselves about when we try to make inferences about mechanism. Sensitivity analysis is something that we do when we have some things that we know we can't resolve in our data, but we can at least make some assumptions and see how far we can be going wrong. You started with the wonderful example of genetics and depression. Everyone says that there is no family environmental effect on the risk of depression. This is strange, because trauma history is strongly predictive of risk of depression, and trauma is strongly familial, but there's no family environmental effect: how can this be? Let's look at meta-analysis. Pat Sullivan did a nice meta-analysis of depression (Sullivan et al 2000) and said that genetics give zero estimates for shared environments. If you go to the literature on the genetics of depression, you get a series of zeros lined up. A statistician would tell you that these estimates shouldn't all be coming out at zero across studies unless there is some sort of confounding going on. Yes, of course, in the twin literature there is confounding of genetic non-additivity and shared environmental effects. Genetic non-additivity is already masking these shared environmental effects. If you do a sensitivity analysis, and make different assumptions about the ratio of additive to non-additive genetic effects, family environment could be explaining 10–20% of the variance in risk without any sort of concerns about interaction. But we have to work with the statistical model and think about what the implications are. You came up with an example of 3 Gs, 1 E; 2 Es, 2 Gs: this is interesting because you are now talking about a mixture of distributions. If there is a binary outcome, this is difficult to test, but if you believe there are quantitative indices of risk, then you should be able to see this. Your Bayesian methods will allow you to ask whether or not you really have a mixture of distributions. Traditionally, in the behavioral genetics literature people haven't done that. They have assumed the same old distribution. The statistical model is important because it helps us to think about what we have in our data. It helps us to look for things that might be telling us that our assumptions are wrong. The statistical model for interaction is interesting. That latent variable interaction term makes some predictions about what should we see in separated twin pairs as opposed to twin pairs reared together. There should be differences in predicted correlations and variances. For personality variables, those differences don't seem to be there. This is not a strong test, but it guides us to look at certain aspects of the data. Don't give up quickly on statistical models, because they can help us think rigorously. You make a good point when you say we should be doing a sensitivity analysis when we are setting these interaction terms to zero. What if we have this assumption wrong? Keeping the statistical modeling framework helps us think this through rigorously.

Uher: I am certainly not trying to jettison statistical modeling. I am suggesting alternatives and complementary strategies that can be used in all the cases where we don't have sufficient power to test the interaction of interest, or the causative mechanism of interest in the available epidemiological samples, because of limitations due to sample composition, quality of measurement and distribution of the genetic and environmental effects. There will be instances where we have a good grounding for expecting interaction effects or a main effect, where we still don't have the means of testing this in a single epidemiological study. I propose alternative ways of exploring these causal mechanisms. One alternative is bringing them into the laboratory where we can randomly assign exposures independently of genes. The other alternative is synthesizing the knowledge across studies. What I intended to demonstrate by the simulations is that the instances where the power is extremely low are very common. There is only a small window of opportunity in which an effect can be detected reliably. I think sensitivity analysis is a good approach for exploring the validity of our assumptions, but it will suffers from the same limitation in power. Often, in sensitivity analysis we look at differential effects of the model, whether we explain the data better or worse. The power of the sensitivity analysis to detect these differences is low. But it is a valid approach.

Heath: My second theme is, let's put these questions in the era of genome-wide association studies. This is where we are now. When we look in that context, you might want to reconsider your assumptions about research design. If people have thousands of cases for genome-wide association studies, the ability to do case-only analysis to look at possible interactions and/or environmental correlations becomes attractive. In a sense, I am not sure it matters in that first step whether the thing that is biologically interesting is the interaction or the correlation. Finding the evidence that either is occurring will then guide us to go further to try to clarify what we have found, but I wouldn't dismiss this as a limitation of the case-only. It is a first step that can take us further.

Uher: I disagree with you about the value of case-only studies. This is for one reason: it is not only that they cannot test their assumptions and therefore lead to spurious findings, but also because they cannot explore the full causal mechanisms and therefore can mask a real, positive finding. It is not just a more powerful design that will get all the positives in, and then we triage the negatives out in a more controlled design, but it also won't get all the positives in. Because of the nature of the statistical term, it will also depend on the sample composition and the measurement. Statistical interactions can be easily transformed into main effects simply by changing the proportion of people exposed. I don't think it is as valuable having a design that is unable to test its own assumptions, and also unable to test main effects. This is my reservation.

Heath: Let us think of it in concrete terms. We are all used to the idea of large cohort studies and prospective studies, on the one hand. At the same time we are hearing that for genome-wide association studies, if we don't have 2000, 5000 or 10000 cases we may be unable to detect effects that are there. 10000 cases for depression is probably not a difficult number to achieve. Think of the Virginia Twin Study, Australian Twin studies, Vietnam-Era Twin studies. All of these samples have history of trauma exposure. Suddenly, I am moving quickly to being able to ask empirically, if I do a case-only analysis, do my trauma exposure people look different? If I had the funding I could do this by the middle of next year. I don't have to wait around for a cohort study. Because early trauma is probably important, I might have to wait 15 years to get results in a cohort study.

Rutter: The points you are making about sensitivity analysis and causal modeling are not in contradiction to the rather separate issue of sample size and the need to combine different designs. James Robins (2001) argued strongly for the need for causal graphs that spell out the statistical implications if the model is working in the way that is proposed, and he has developed methods that can take this forward. Similarly, epidemiologists have been pressing for quite a long time on the need for using sensitivity analyses (see Susser et al 2006). Indeed, one of the points made a long time ago in relation to smoking and lung cancer was that the sensitivity analyses made it clear that a confounding effect would have to have at least nine times the effect of smoking in order to account for the smoking association (Cornfield et al 1959). No one has been able to think of a confounder that would do this. Equally, there other examples where sensitivity analyses have landed up with an entirely different conclusion. I hope we don't go down the route of saying that we don't need statistics. To the contrary, we fundamentally need statistics: the question is, how do we use them, and how can we provide ways of thinking of the alternatives? Cochran & Chambers (1965) many years ago argued that the issue most overlooked in looking for causal influences is postulating that if it is not this, what might it be, and what statistical analysis and design is needed to tease these apart? This has to be fundamental.

Heath: The theme I am trying to add to the discussion is the question of whether we can take advantage of the G × E interaction for gene discovery, where we don't have a clear mechanism. There, the case-only design will be the one that advances us the most rapidly.

Dodge: I am so taken by what we have heard so far that I want to come to a conclusion and a prediction. The conclusion is that every effect is a G × E interaction. There is no such thing as a gene main effect or an environment main effect. It is a theoretical impossibility because genes are contextualized in environments, and environments are contextualized with genes. Michael Rutter made the excellent point that there is a huge difference between the importance of a factor and the importance of variation in that factor. That is, a factor—gene or environment—may be hugely important, but our sample or the environment might not have sufficient variation to test it. A second point is that what we know about evolution is that across history and time, genes have shaped environments and vice versa. That is, the gene and environment constellation today is a function of natural selection, and they have selected each other. We have evolved given this environment, and the theoretical variation in genes and the theoretical variation in the environment is much larger than exists today. It is a theory experiment to say that there is always going to be a G × E interaction, because I can always contemplate an environment that would demonstrate an environmental effect, even though the environment doesn't exist today. Likewise, some genes have been selected out and don't exist today. So my prediction is that we will do these G × E interaction studies for the next 10 years, and we will be excited by the discoveries, but after 10 years we will stop. I think the point will be proven that it is all G × E. Then I think we will start doing the other kinds of studies—case and experiment studies—and move to some theory about what a gene does and what an environment does. We need to move beyond G × E to theory, mechanisms and processes to understand how disorder develops.

Heath: I have a final question. Whose genome anyway? In other words, depending on the phenotype that you are looking at, the microbial communities that live inside us show genetic variation, and have the potential to affect the products of the foods we eat. They also can be interacting with our dietary environment. We are in the era of the human microbiome project. Genotype is not necessarily our genotype: it may be the genotype of the microbes that have co-evolved with us.

Poulton: Andrew Heath, if I understand your point, you are saying that statistical approaches are useful tools to give us a heads-up about what we are not seeing or where we might be going wrong. But you are not saying that statistical interactions should define concepts which are essentially biological.

Heath: That's correct.

References

Cochran WG, Chambers SP 1965 The planning of observational studies of human populations. J R Stat Soc [Ser A] 128:234–266

Cornfield J, Haenszel W, Hammond EC, Lilienfeld AM, Shimkin MB, Wynder EL 1959 Smoking and lung cancer: recent evidence and a discussion of some questions. J Natl Cancer Inst 22:173–203

Robins JM 2001 Data, design and background knowledge in etiologic inference. Epidemiology 11:313–320

Sullivan PF, Neale MC, Kendler KS 2000 Genetic epidemiology of major depression: review and meta-analysis. Am J Psychiatry 157:1552–1562

Susser E, Schwartz S, Morabia A, Bromet EJ 2006 Psychiatric epidemiology: searching for the causes of mental disorders. Oxford University Press, Oxford & New York

3. Gene–environment interaction and behavioral disorders: a developmental perspective based on endophenotypes

Marco Battaglia, Cecilia Marino*, Michel Maziade†, Massimo Molteni* and Francesca D'Amato‡

*The Department of Psychology, 'Vita-Salute' San Raffaele University, 20127 Milan, Italy, * The Department of Child Psychiatry, Istituto Scientifico Eugenio Medea, 23842 Bosisio Parini, Italy, † The Department of Psychiatry, CRULRG, Laval University, Québec, Canada G1K7P4 and ‡ The Department of Psychobiology & Psychopharmacology, CNR Institute of Neuroscience, 00143 Rome, Italy*

Abstract. It has been observed that '*No aspect of human behavioral genetics has caused more confusion and generated more obscurantism than the analysis and interpretation of various types of non-additivity and non-independence of gene and environmental action and interaction*' (Eaves LJ et al 1977 Br J Math Stat Psychol 30:1–42). On the other hand, a bulk of newly published studies appear to speak in favour of common and frequent interplay—and possibly interaction—between identified genetic polymorphisms and specified environmental variables in shaping behavior and behavioral disorders. Considerable interest has arisen from the introduction of putative functional 'endophenotypes' which would represent a more proximate biological link to genes, as well as an obligatory intermediate of behavior. While explicit criteria to identify valid endophenotypes have been offered, a number of new 'alternative phenotypes' are now being proposed as possible 'endophenotypes' for behavioral and psychiatric genetics research, sometimes with less than optimal stringency. Nonetheless, we suggest that some endophenotypes can be helpful in investigating several instances of gene-environment interactions and be employed as additional tools to reduce the risk for spurious results in this controversial area.

2008 Genetic effects on environmental vulnerability to disease. Wiley, Chichester (Novartis Foundation Symposium) p 31–47

The clear evidence that most—if not all—complex behavioral traits, including the categories of mental disorders encompassed by current classification systems such as the DSM-IV or the ICD-10, have a multi-factorial etiology has greatly helped to foster the notion that common illnesses such as depression or panic disorder are by many points of view similar to diabetes or cardiovascular disorders, hopefully helping to reduce stigma against mental illness.

Some three decades of increasingly sophisticated biometric behavioral genetic research and modelling applied to a host of psycho(patho)logical conditions have shown quite convincingly how it can be safely assumed that underlying these conditions is a normally distributed multifactorial liability, whereby several genetic and environmental causative agents—each providing a small individual contribution to the observable risk for illness—act additively, independently and in a probabilistic, not deterministic, fashion. Moreover, item-response models applied to large epidemiological samples screened for depression or sociopathy in the developmental age have shown that even in presence of strongly skewed phenotypic distributions, a normally distributed underlying liability can be assumed quite safely (van den Oord et al 2003) for these phenotypes.

When the influences of specific genetic variants are analysed in the context of environmental moderators, however, the findings often show small—but significant—average genetic effects at the population level, and considerable variability of effect size across different strata of environmental risk factors (Nobile et al 2007). Consistent with these observations, a clear understanding of the interplay between genes and environment in shaping individual risks for psychopathology has been set amongst the goals of contemporary behavioral genetic studies (Rutter et al 2006, Moffitt et al 2005, Eaves 2006).

Gene–environment interactions in behavioral and psychiatric genetics: data and views from different vantage points

At present, two different approaches to gene–environment (G × E) interactions offer an interesting opportunity to evaluate quite divergent results and implications.

On the one hand is the classical behavioral genetics biometric approach: by this method strong G × E interaction has rarely been found. In a recent work Eaves (2006) begins his critical reappraisal of G × E interactions in behavioral disorders by noticing that even though interactions appear to be widespread in several experimental organisms, their contribution is typically smaller than those of main effects (Mather & Jinks 1982). Moreover, G × E interactions can spuriously emerge—or be removed—from results in any typical human behavioral genetic study by the simple act of transforming a variable, as routinely conducted in the attempt to treat heteroscedasticity. Thus, by paying insufficient attention to scale of measurement problems one may obtain inaccurate estimations of genetic and environmental effects, and—of specific relevance to the main topic of this paper— embrace unwarranted explanations implying G × E interactions when simpler models for the effects of genes and environment are actually more accurate and realistic (Eaves et al 1989).

In order to investigate the possibility that some—or several—claims of recent G × E interactions in human behavior and psychopathology can actually constitute artefacts, Eaves (2006) conducted simulations under a model that allowed for mixture of distributions in liability conditional on genotype and environment. Multiple blocks of simulated data were analyzed by standard statistical methods to test for the main effects and interactions of genes and environment on liabilities and diagnoses of major depression and antisocial behavior. Analysis of the dichotomized data by logistic regression frequently detected significant G × E interaction but none was present for liability. The author concluded that the biological significance of some published findings of G × E interactions can be questionable (Eaves 2006).

An alternative view of, and approach to G × E interaction is well summarized in a series of recent conceptual reviews by Michael Rutter and associates. Rutter and colleagues (Rutter et al 2006) maintain that the dismissal of gene–environment interaction in human behavioral genetics in the eighties and early nineties arose from quantitative genetic 'black box' analyses of anonymous (i.e. unmeasured) G and anonymous E, which usually failed to show G × E (Plomin et al 1988). Rutter and associates emphasize that what was being tested for in that context was an omnibus interaction between all genes and all environments, which would be biologically unlikely, so that almost invariably G × E interactions would not be identified.

Consistently, a clear differentiation has been set between *theoretical* G × E interactions and *measured* G × E interactions, the latter literally representing the possible interaction between one or more identified environmental risk factors and one or more identified gene variants (Rutter et al 2006, Moffitt et al 2006).

Sometimes G × E interactions have first been successfully simulated in biometric models, and then confirmed in empirical research based upon identified functional genetic polymorphisms and identified environmental risk factors. An interesting example of coherent transition from theoretical models to empirical findings in human behavior is provided by applications to juvenile antisocial behaviors and environmental mediation of the genetic risk, whereby the simulations by Turkheimer & Gottesman (1991) based upon tridimensional reaction surfaces which suggested non-linear phenomena and likely gene-by-environment interactions (Goldsmith & Gottesman 1996) received apparent confirmation by a study of the role of genotype in the cycle of violence in maltreated children (Caspi et al 2002) in the Dunedin longitudinal cohort.

Nonetheless, there seems to be a clashing contrast between the sobriety suggested by the results of rigorous behavioral genetic model simulations and the enthusiasm fuelled by an increasing number of claims of G × E interactions available in the literature.

Since factors such as study design, sample size and genotyping technology can influence the analysis and interpretation of observed interactions, some of the current major issues in the field are the identification of robust strategies to conduct fruitful studies of G × E interactions (Hunter 2005) and ascertain the validity of the findings (Moffitt et al 2006).

G × E interactions in behavioral and psychiatric genetics: the risk for artefacts and possible remedial strategies

Moffitt and associates (2006) have suggested a series of possible remedies to avoid false findings of G × E interactions. They articulate their proposals for ascertaining the validity of G × E findings upon three main points summarized in Table 1. The first point addresses how to deal generally with scaling effects and artefacts. Under point 1 they suggest three practical approaches: (a) potential scaling artefacts can be circumvented by substituting for the genotype of interest a similarly distributed polymorphism without theoretical relation to the hypothesis under study. For instance, in the Caspi et al (2002) study on MAOA and conduct problems, they analyzed the possible effect of a random single nucleotide polymorphism with allele frequencies similar to the MAOA, and found no evidence of interaction with maltreatment in predicting conduct problems. Following a similar but speculative approach, they showed that: (b) by picking a disorder with the same prevalence as conduct disorder—gum disease—and checking whether it was predicted by the interaction of MAOA and maltreatment, they found that it did not. Along this same line of reasoning Moffitt et al (2006) suggest a third way of

TABLE 1 Moffitt et al's (2006) approach to some problems in G × E research

Problem	Possible remedies
Scaling effects and artefacts, general	a) <u>Genotype-wise</u>: try similarly distributed polymorphisms w/o relation to hypothesis under study; b) <u>Phenotype-wise</u>: i) try similarly distributed phenotypes w/o relation to hypothesis under study; ii) sensitivity analysis: see if G × E holds across different phenotypic measures that map the same diagnostic construct.
Scaling effects and artefacts in omnibus multiplicative tests	Planned group comparisons—whenever sound theory and knowledge is available to break down hypothesis into a-priori defined groups of risk/resilience- instead of overall multivariate G × E analysis via general interaction term.
Biologically meaningless/flawed G × E finding	Nomological network evidence that: a) the environmental influences biological systems involved in disorder; b) the candidate gene is associated with animals' and humans' reactivity to the environmental pathogen; c) the putative risk factor has true environmental causal effects on the disorder.

probing under point 1: (c) if the G × E interactions reflects a valid biological process, then it ought to robustly predict the same disorder outcome measured in different ways, each having its own metric properties. Consistently in the Caspi et al (2002) paper it was demonstrated that a G × E interaction between MAOA and violence predicted a host of different measures of dissocial functioning—all sharing the same construct validity—such as a categorical diagnostic measure (e.g. conduct disorder), a scale of symptoms (e.g. number of conduct problems), a personality trait scale (e.g. aggressive personality), an informant's rating (e.g. antisocial lifestyle), or an official record (e.g. criminal conviction). This latter strategy, however, may appear somehow weaker, in that it is subject to a certain methodological circularity (all these 'alternative' phenotypic measures are necessarily strongly correlated), and most importantly, it has been shown (Eaves 2006) that even with categorical data and logistic regression the danger of finding artefactual G × E effects persists. On the other hand, having an interaction confirmed by different measurements of the same constructs can be reassuring, since it could well be conceived a scenario where the extension of comparisons to related—but different—phenotypes, could yield different patterns of interaction, or no interactions at all.

The second anchor point recommended by Moffitt et al (2006) to reduce spurious interactions is to achieve a good match between the predictions derived from the G × E hypothesis and the statistical approach used to operationalize it. Briefly, rather than relying on putting an interaction term into an overall multivariate analysis (Rutter & Pickles 1991) which would in some cases allow a false statistical main effect of genotype to absorb part of the variance of the interaction, they recommend planned-in-advance group comparisons, based upon clear 'a priori' theoretical information and hypothesis about the genotype and environment. This would help to break down the G × E hypothesis to predict the precise pattern of psychopathology. The third point is more general, and much more connected to biology as an empirical discipline: it is indeed an incitement to explore directly and deliberately G × E interaction by *following a biologically plausible theoretical rationale behind the hypothesis including possible associations between specified genes and animals' and humans' reactivity to the environmental pathogens*' (Moffitt et al 2006).

Endophenotypes: can they be valuably added to the toolbox for adopting or discarding new findings of G × E interaction?

Regardless of the level of cautiousness towards recent reports of G × E interactions, several authors with different orientation (Eaves 2006, Moffitt et al 2006, Hunter 2005) seem to agree that a bottleneck—both conceptual and methodological—for adopting or discarding new findings lies in the measurable, ultimate biological value and meaning of any new finding in this specific area.

We suggest that carefully chosen endophenotypes can represent an additional, possible tool to investigate G × E interplay/interaction in a relatively stringent multilevel approach, as they can be of some use in preventing false positive results and foster heuristic research in behavioral and psychiatric genetics.

The original use and meaning of the term endophenotype as coined by Gottesman & Shields (1972) was specific, in that it described a measurable trait which is at the same time distinct from the diagnosis and which is occurring more commonly in both affected and unaffected family members than in the general population. However, the frequency of use of the term 'endophenotype' has recently increased dramatically in the behavioral and psychiatric literature, so that it is often employed for a host of traits that would better and more simply qualify for biological markers (see Gottesman & Gould 2003 for a discussion of the differences between these two concepts and measures). The 'endopheno-type' term is also used occasionally to describe quantitative phenotypes that occur only in affected family members. Recently Waldman (2005) proposed a working set of 10 criteria (summarized in Table 2) to define valuable endophe-notypes and outlined analytic methods for evaluating their validity and utility in a developmental behavioral genetic perspective. Interestingly, in his proposal of neuropsychological tests (Trail A and B) as putative endophenotypes applied to ADHD and to the DRD4 genotype, Waldman (2005) demonstrated that empirical probing satisfies most—but not all—of the 10 criteria. A very recent paper by Szatmari et al (2007) offers a critical, helpful review of the use and application of endophenotypes to psychiatric genetics, while Hariri & Weinberger (2003) approached related issues, albeit within an imaging genomics framework of reference.

Suppose now that a valuable endophenotype is identified for a given behavioral disorder (dimensionally or categorically defined): by virtue of it closer proximity to the genetic mechanisms underlying the disorder, the endophenotype is likely to be more heritable than the disorder itself (Battaglia et al 2007a), but inevitably it will encompass a causal component which will be attributed to the environment, be it shared or idiosyncratic in nature. Likewise, we can well expect that a classical design that addresses the origin of covariation between the disorder and the endo-phenotype—such as a bivariate twin study—will yield results of significant genetic covariation. But environmental mediation can be heuristically important and fruit-ful, since both genetic and environmental causal covariation between a phenotype and an endophenotype will indicate that partially shared causal agents have an effect upon the clinical phenotype that we want to study, and a process indexing a corresponding neural substrate. Of course, by classical behavioral genetics approaches one is simply assessing the overlap between the causal agents, or the nature of covariation, between a phenotype and a corresponding endophenotype. Such shared causal agents could still be acting entirely linearly and additively:

TABLE 2   Waldman's (2005) ten criteria for the selection of valid endophenotypes

1) e. has appropriate psychometric properties; (reliability and construct validity; test for normality, skewness, and kurtosis.)

2) e. is related to d. and its symptoms in the general population (variance of e. can be substantially restricted in clinical samples ascertained for presence of d.; thus: examine the relationship of putative e. to symptoms of d. in general population, where variance might be maximized; possibly analyse by teterachoric correlations)

3) e. is stable over time (i.e., is expressed regardless of whether or not the disorder is currently manifest; may be assessed by longitudinal assessment of predisposed individuals);

4) e. is expressed at a higher rate in the unaffected relatives of probands than in randomly selected individuals from the general population (family studies);

5) d. and e. co-segregate within families (sibling analyses);

6) e. is heritable (classical twin study, univariate with adjustment for possible selective ascertainment)

7) there are common genetic influences underlying the endophenotype and the disorder (twin bivariate study)

8) e. shows association and/or linkage with one (or more) of the candidate genes or genetic loci that underlie d., and should show association with the gene over and above the gene's association with d. or its symptoms (multiple molecular genetic strategies available)

9) e. should *mediate* association and/or linkage between the candidate gene and d., meaning that the effects of a particular gene or locus are expressed—either in full or in part—through the endophenotype

10) e. should *moderate* association and/or linkage between the candidate gene and d., meaning that the effects of a particular gene or locus on a disorder are stronger in disordered individuals who also show the endophenotype.

In parenthesis are reported some of the strategies that can be adopted to investigate/validate criteria. e, endophenotype; d, disorder; modified by Waldman 2005

possible G × E may remain unmeasured and/or undetected because of limited statistical power or inadequate modelling (Rutter et al 2001, 2006).

However, inasmuch as the same endophenotype can be exported into an animal model, a series of rigorous tests of identified genetic and environmental causal agents, alone and in interactions, would become possible in the context of hypotheses formulated explicitly and *a priori* (Battaglia & Ogliari 2005) and be extensively explored in the animal before they can be investigated in humans.

Application to a developmental model of human panic

As an example, we are currently working on a program that interfaces human and animal liability to overreact to carbon dioxide (CO_2) inhalation and the risk for developing panic disorder (PD) in genetically informative studies, including twin studies, and an animal model of sensitivity to CO_2 in response to early separation from the mother. By capitalizing on the unconditioned response to CO_2 in a sample of twins from the Norwegian general population, we (Battaglia et al 2007a) were able to show that: (1) an exaggerated (i.e. overanxious, and/or hyperventilatory) response to a CO_2 stimulation challenge is about 30–50% more heritable than the average indexes reported for the clinical phenotype of PD; (b) milder and stronger responses to the challenge lie along the same continuum of liability; and (c) that according to a full model corrected for the partially selective ascertainment, the phenotypic correlation between post-CO_2 anxiety and DSM-IV panic is largely due to additive genetic influences with shared and unique environmental agents concurring to explain a relatively minor proportion of the correlation between these two traits, while the best-fitting model showed a genetic correlation between post-CO_2 anxiety and panic of 0.81 (95%CI: 0.50–0.98, Battaglia et al 2007b). By the criteria defined above, CO_2 responsiveness appears to constitute a reasonable endophenotype of human panic.

Interestingly, the hyperventilatory and anxious responses to the hypercapnic stimulus are an unconditioned, adaptive trait common to all mammals, and the responses to heightened CO_2 concentrations can be easily and non-inferentially transferred to the mouse by relatively simple plethysmographic measurements, and standard behavioural tests. The laboratory allows for rigorous control of environmental variables that can be studied alone, or in combination with identified genetic variables.

By capitalizing on the notions that separation anxiety and early separation from/loss of parents influence the individual disposition to panic disorder (Kendler et al 1992, Battaglia et al 1995) and to CO_2 reactivity (Pine et al 2000), we are now comparing mouse pups that have been exposed to separation from their mother and normally reared pups for their responsiveness to an unconditioned CO_2 challenge. Our preliminary, unpublished results show a dramatic increase of several ventilatory parameters in response to hypercapnia—but not to hypoxia—in the former group, which shows impressive stability of the trait once animals have reached adulthood, and a parallel tendency to show an excess of avoidant behaviours in classical tests of unconditioned avoidance.

Since some neurotransmitters' systems—such as the serotonergic, cholinergic and the opioid systems—are simultaneously implicated in mediating chemoception, respiratory physiology, early attachment/affiliative behaviours and anxiety, they would appear to constitute logical targets in the context of molecular genetics approaches to these issues. Most importantly, the molecular genetic experimentations that will follow for this animal model of human panic, will be able to explore possible G × E interaction in a rigorously controlled environment, consistent with *a priori* outlined hypotheses formulated by our group (Battaglia & Ogliari 2005), and will guide the successive, possible extensions to humans.

An ideal experimental agenda within this framework of reference would then be:

(a) an endophenotype bearing a biologically meaningful, theoretical rationale is found to covary with a clinical phenotype because of shared genetic liability between the two conditions;
(b) the endophenotype can be plausibly exported to animal experimentation with minimal inferential reading;
(c) within the context of the animal model one or more polymorphisms, and one or more environmental risk factors are found to be associated with individual variation of the endophenotypic response;
(d) there is ground to sustain significant G × E interaction within in the animal model, which can be probed and replicated at multiple levels (microarray, micro-PET, changes in alternative splicing, modifications of endophenotypic response by siRNA-mediated suppression of gene levels and transgenic over-expressing of specific genes, etc.)
(e) the same polymorphism is now investigated in human, both at the endophenotype and the clinical phenotype levels, both linearly and in possible G × E interactions. If association and direction of possible G × E effects are found to be the same as in the animal model, this consistence is taken as an indicator of validity.

By this 'from human to animal and back' agenda (Battaglia & Ogliari 2005) one may decide that only after having rigorously explored in the carefully controlled context of the endophenotype exported to the animal model the possible genetic, environmental and G × E effects, will carry on exploration of the same causal chains in human samples.

Needless to say, there are several drawbacks in this approach. First, there are indeed few endophenotypes that can be easily and credibly exported from 'animal to human and back'. Second, this approach is time-consuming and requests a strong collaboration between scientists that work on animals and those working on humans: these two worlds are still much too far apart and there is still too little cross-talking. Third, albeit there is about 80% synteny between humans and mice, the correspondence between polymorphisms in humans and mice remains relatively limited.

Finally, focused as it may be, any endophenotype will encompass multiple biological systems and functional measures, several of which may act as 'latent variables' and show association without major impact on pathology. For instance, in our specific approach we may find effects for genes and/or environmental agents that influence respiratory variables, and much less emotionality or panic.

With these limitations in mind we suggest that adopting valid endophenotypes and exporting them to animal experimentation can substantially improve our comprehension of the genetic and environmental causes that, alone and in interaction, shape individual risks to psychopathology along the lifespan.

References

Battaglia M, Ogliari A 2005 Anxiety and panic: from human studies to animal research and back. Neurosci Biobehav Rev 29:169–179

Battaglia M, Bertella S, Politi E et al 1995 Age at onset of panic disorder: influence of familial liability to the disease and of childhood separation anxiety disorder. Am J Psychiatry 152:1362–1364

Battaglia M, Ogliari A, Harris J et al 2007a A genetic study of the acute anxious response to carbon dioxide stimulation in man. J Psychiat Res 41:906–917

Battaglia M, Pesenti-Gritti P, Spatola CAM, Ogliari A, Tambs K 2007b A twin study of the common vulnerability between heightened sensitivity to hypercapnia and panic disorder. Am J Med Genet Part B, in press

Caspi A, McClay J, Moffitt TE et al 2002 Role of genotype in the cycle of violence in maltreated children. Science 297:851–853

Eaves L J 2006 Genotype × Environment interaction in psychopathology: Fact or artifact? Twin Res Hum Genet 9:1–8

Eaves LJ, Last KS, Martin NG, Jinks JL 1977 A progressive approach to non-additivity and genotype–environmental covariance in the analysis of human differences. Br J Math Stat Psychol 30:1–42

Eaves LJ, Eysenck HJ, Martin NG 1989 Genes, culture and personality: an empirical approach. New York: Academic Press

Goldsmith HH, Gottesman I 1996 Heritable variability and variable heritability in developmental psychopathology. In: Lenzenweger MF, Haugaard JJ (eds) Frontiers of developmental psychopathology. Oxford University Press, New York, p 5–43

Gottesman II, Shields J 1972 Schizophrenia and genetics: a twin study vantage point. New York: Academic Press

Gottesman II, Gould TD 2003 The endophenotype concept in psychiatry: etymology and strategic intentions. Am J Psychiatry 160:636–645

Hariri AR, Weinberger DR 2003 Imaging genomics. Br Med Bull 65:259–270

Hunter DJ 2005 Gene-environment interactions in human diseases. Nat Rev Genet 6:287–298

Kendler KS, Neale MC, Kessler RC, Heath AC, Eaves LJ 1992 Childhood parental loss and adult psychopathology in women. A twin study perspective. Arch Gen Psychiatry 49:109–116

Mather K, Jinks JL 1982 Biometrical genetics: the study of continuous variation (2nd ed.). London:Chapman Hall

Moffitt TE, Caspi A, Rutter M 2006 Measured gene-by-environment interactions in psychopathology. Psychol Sci 1:5–27

Nobile M, Giorda R, Marino et al 2007 Socioeconomic status mediates the genetic contribution of the DRD4 and 5-HTTLPR polymorphisms to externalization in pre-adolescence. Dev Psychopathol 19:1145–1158

Pine DS, Klein RG, Coplan JD, Papp LA, Hoven CW, Martinez J 2000 Differential carbon dioxide sensitivity in childhood anxiety disorders and non ill comparison group. Arch Gen Psychiatry 57:960–967

Plomin R, DeFries JC, Fulker DW 1988 Nature and nurture during infancy and early childhood. New York: Cambridge University Press

Rutter M, Pickles A 1991 Person-environment interactions: concepts, mechanisms, and implications for data analysis. In: Wachs TD, Plomin R (eds) Conceptualization and measurement of organism-environment interaction. American Psychological Association, p 105–141

Rutter M, Pickles A, Murray R, Eaves L 2001 Testing hypotheses on specific environmental causal effects on behavior. Psychol Bull 127:291–324

Rutter M, Moffitt TE, Caspi A 2006 Gene–environment interplay and psychopathology: multiple varieties but real effects. J Child Psychol Psychiatry 47:226–261

Szatmari P, Maziade M, Zwaigenbaum L et al 2007 Informative phenotypes for genetic studies of psychiatric disorders. Am J Med Genet B Neuropsychiatr Genet 144B:581–588

Turkheimer E, Gottesman I 1991 Individual differences and the canalization of behaviour. Dev Psychol 27:18–22

van den Oord EJC, Pickles, A Waldman I 2003 Normal variation and abnormality: an empirical study of the liability distributions underlying depression and delinquency J Child Psychol Psychiatry 44:180–192

Waldman ID 2005 Statistical approaches to complex phenotypes: evaluating neuropsychological endophenotypes for attention-deficit/hyperactivity disorder. Biol Psychiatry 57:1347–1356

DISCUSSION

Heath: With endophenotypes, one of the questions is always the specificity. Likewise, with your mass exposure model there is suppression of specificity of effects. Early separation in humans has a broad range of outcomes, including depression. If you go back to your Norwegian twin registry data and compare panic with comorbid depression, depression only and controls, what sort of pattern to you see in response to CO_2 challenge? Is depression-only also associated?

Battaglia: No, we are controlling for depression. We don't find that depression plays a significant role in affecting response to the CO_2 challenge. I see your point that early separation can really affect human development in many different ways. On the other hand, early separation is affecting panic quite specifically. Kendler et al's (1992) data on twin women's psychopathology showed that the variance in liability for several internalizing diagnoses including depression, panic, generalized anxiety and phobia is affected by childhood parental loss, but the effect was highest for naturally occurring panic, and within panic, inversely related to age at occurrence of loss. Likewise, in our candidate twin data, one of the strongest partially environmental predictors of response to the CO_2 challenge is parental loss. We were careful in investigating this with the same criteria as adopted by Kendler et al, that is a parental loss having occurred before age 17 including death, divorce and departure for the military. But I take your point that it is a big smash against normal development which could and often does affect many outcomes, including depression.

Heath: Do you have history of childhood trauma on that sample? Is it correlated with panic? Is it predictive of CO_2 response? Early parental separation does not occur in a vacuum. If we look at histories of early trauma and parents leaving the home there is an interesting coincidence of timing. There is trauma, and the parent leaves.

Battaglia: I agree, they can often be associated. To start with, the twin cohort for which we had CO_2 is relatively small. It is about 350 pairs, which is large for CO_2 provocation, but small compared with the average size of modern twin studies, and so the power is limited. We have records of several major life events of different severity, including physical assault. In general, the various categories of adverse events happen to be quite weakly reciprocally correlated (between 0.08 and 0.15) in this general population twin sample. Also, in our regression models the effect of every predictor is controlled for the effect of all other predictors. Major life events—such as serious illnesses or physical assault—seem to exert a smaller influence than early parental loss on both naturally occurring panic and on CO_2 challenge response. But again, this might be related to the relatively small sample size, and severe events such as physical assaults in childhood are rare.

Heath: Do you have a hypothesized mechanism by which you get from early separation to altered response when you are genetically vulnerable?

Battaglia: Yes, one mechanism might have to do with the cholinergic system: intense stress causes some alternative splicing of acetylcholine esterase (Kaufer et al 1998). This has been found to be protective for the brain. One of our hypotheses however is that the same mechanism which may be protective for the higher brain can be a risk factor for the lower brain, for instance the medulla, because it may enhance sensitivity to suffocatory stimuli (Battaglia & Ogliari 2005). This is a hypothesis based on careful animal observation, and one thing we will do in the future is to look at mice overexpressing acetylcholine esterase. Other reasonable candidates along this causal pathway include genes related to the serotonergic and the opioid systems, as they are both simultaneously implicated in respiration, anxiety and affiliative behaviors.

Kleeberger: About 10 years ago when I was still at Johns Hopkins, we examined interstrain variation among inbred mice and their response to hypercapnia, hypoxia and normoxia challenges. We found significant differences between strains in terms of the inspiratory and expiratory timing, their breathing frequency and tidal volumes. Dr Clarke Tankersley at Hopkins has since continued these studies and has found some strong quantitative trait loci (QTLs) that contribute to the response to hypercapnia (Schneider et al 2003, Tankersley & Broman 2004). He is not a behavioral biologist and has not looked at that phenotype, but it seems that the kind of work you are doing and the kind of work he is doing should be integrated. These genes that control the physiology of the response could have an important impact on the behavior aspects that you are investigating.

Battaglia: I didn't know about that work. Of course, we have to be careful to take into account possible background effects related to specific strains. This is also why we are working on outbred mice at this time.

Kleeberger: From a gene discovery standpoint, there are a number of existing and emerging tools that could assist in parsing out some of these behavioral phenotypes.

Kotb: When you did your studies with outbred mice, taking one member of the litter and weaning it with a foster mother, have you done a study where you waited different times before separating the mice from their mother? And did you see any difference?

Battaglia: That's a good question, and it is something we are now investigating. At present it would appear that even one cross-fostering phase (i.e. a passage to a foster mother after 24 h rearing by biological mother) would be enough to influence several key respiratory parameters in these pups. But this is truly a preliminary hint, we need more data on this.

Kotb: In addition to the psychological aspect, I'm thinking of immunological aspects: you are taking these babies from one mother and putting them with another mother who has different immune exposure and a different immune system.

Battaglia: Francesca D'Amato who I am collaborating with, and who designed the cross fostering paradigm is taking care of these aspects carefully. This is all I can tell you.

Kotb: You can do this with recombinant inbred mice: you can use a foster mother from the same strain versus a foster mother from a different strain, this will allow to determine whether weaning the litter with genetically identical vs. genetically distinct foster mothers makes a difference—either way it would be very interesting.

Martinez: Could these be epigenetic effects? Could methylation of candidate genes be involved? And could these be passed on to the next generation? It might be interesting to take these anxiety mice and determine whether their offspring are more prone to anxiety. It would be an interesting area to explore.

Battaglia: I agree with both of you. Indeed, applying the paradigm to Jax recombinant mice strains would help finding which DNA areas are associated with the response, and of course the epigenetic aspect is worth investigating.

Braithwaite: You apply a trauma early on, which is separation in the case of these mice. What is the 'memory' that allows you to get this differential response compared with controls at 18 d and 60 d? I noticed in one of your experiments the basal level of the unstressed animals that had been fostered was higher than in the controls. This suggests that there was already an anxious response occurring. Have you looked over time in the fostered animals to see whether the basal levels of the cholinergic neurotransmitters are actually higher than in your control animals?

Battaglia: This is a good question. It could partially be addressed by the strategy Malak Kotb is suggesting, i.e. going back to look at what happens with shorter separations and see whether we get some kind of dose effect in the readouts.

Another suggestion comes from human studies. In cohorts from us and others, surprisingly the correlation between separation anxiety and actual separation events is weak, in the order of 10%. It is not that one condition necessarily implies the other, in humans at least. There is a weak correlation, but ideally one would think that there are at least two partially independent pathways to get to this hypersensitivity to hypercapnia in humans: being very anxious about separating from your parents without a real separation event in view (the definition of separation anxiety), or true separation and getting to a series of complex situations including panic anxiety. In humans the correlation is weak.

Braithwaite: There still has to be a molecular memory, somehow. There has to be something that is different in those traumatized animals. If you have some evidence that the neurotransmitters are expressed at a higher level, which would be something that may be maintained over a lengthy period.

Rutter: Can I ask you to say a bit more about endophenotypes? The example you gave was a persuasive one: clearly, it worked in a useful way, and provided a means of bridging between animal studies and human studies. There are concerns about the argument that endophenotypes have to be closer to the gene. This seems to me to be a wrong-headed assumption. For example, in looking at the effect of drugs on AIDS, T cell responses have been used as a way of looking at a feature on the causal pathway, which is why it is of interest. It has nothing to do with there being a main effect of genes. The evidence in psychopathology that the so-called endophenotypes that have been used are more heritable, is mainly negative. Yours is a counter example. Do you see this as a general solution in the way that Irv Gottesman would like us to believe?

Battaglia: I think the word 'endophenotype' is overused in psychiatry. It's a sexy word that everyone seems to love nowadays. Then you read a paper and wonder why people have used the term 'endophenotype' to describe what is simply another behavioral measure. When I was in my psychiatry years, almost no one used this word. Gottesman had coined the term but almost no one used it: instead they talked about biological markers. Paradoxically, some people said that the best biological marker is one that has nothing to do with the causal pathway leading to mental illness. I never understood why, but this was one of the golden rules at the time. Now people say the opposite. Many also think it must be more heritable that the corresponding clinical phenotype. I don't think this is that important. It is important to show that at least some of the covariation is due to sharing of some genes. That it is more heritable isn't vital, in my opinion. The important thing is that they go together causally, ideally sharing one or more genes. Finally, to me, as a rule of thumb to prove robustness, you should be able to find that the correlation goes through all three levels: gene, intermediate phenotype and behavior. In other words: A predicts B, B predicts C, and A predicts C, which is almost never the case if you read carefully the papers in the field. We were happy when we saw such good fit in our ERP paper of the responses to facial expressions in children with different degrees of with social anxiety (Battaglia et al 2005). We used the serotonin transporter, ERP and social anxiety, and we were happy to show that all the three levels go well together. In many papers, even in some very well-published ones, regressions or correlations work at two levels, A with B and A with C. The ideal situation is to show that the relationship goes through the gene, to behavior, passing through an intermediate phenotype. Strong heritability can be there or not.

Rutter: Neither you or Andrew Heath used the term co-morbidity, but the discussion you had was related to the extent to which effects are specific. Specificity was one of the things that Bradford Hill put forward in his nine guidelines, but in the review that I chaired a working party on for the Academy of Medical Sciences (2007) we agreed with epidemiologists who said that is actually something that should be jettisoned (Rothman & Greenland 1998). This is because both environmental effects and genetic effects so frequently have multiple non-specific outcomes. On the other hand, if you have specificity it helps. So where do we go?

Heath: My take on this is that there are many things that are not specific, but it is always important to check whether we do have specificity. As an alcohol researcher who had a history in smoking research, I have taken great pleasure over the years in watching the community of smoking researchers ignore the fact that quite a high proportion of smokers have a history of alcohol problems, and the community of alcohol researchers ignoring the fact that a very high proportion of people with alcohol problems are also smokers. Therefore they completely misspecify some of the risk factor associations that they were discovering. Often effects are non-specific, but it is important that we always look to see whether they are specific and specific for something that we weren't imagining when we set out to do our research.

Rutter: I agree.

Heath: Another example of an endophenotype is within the alcohol field, with alcohol challenge and ADH genes. For these genes we see differences in alcohol metabolism and response to alcohol, we hypothesize, is an important component of what determines risk for problems with alcohol. Then it makes sense to go after that particular endophenotype or intermediate phenotype. The Australian alcohol challenge studies have been quite helpful in showing genetic association of genes in the ADH gene region.

Martin: You don't become an alcoholic if you find drinking alcohol so unpleasant.

Heath: My point is that sometimes endophenotypes do work.

Rutter: Yes, they do. It is just that one has to be careful not to see them as a total answer. And certainly not confined to getting closer to the genes. That is very useful for some things, but in the field of internal medicine its uses go much broader than that.

Poulton: The display of how endophenotypes advance understanding in animal models is fantastic and convincing. But I think the use of endophenotypes in human populations can distract attention from a key issue which is basically the validity of the phenotypes that psychiatrists currently use. The work that we (Gregory et al 2007) and David Fergusson (Fergusson et al 2006) have done recently suggests that for internalizing disorders there is probably a higher order trait that can be broken down into the lower order, correlated facets of distress and fear. You said that you thought that the term endophenotype was a bit of a fashion: is what lies behind that comment a cynicism about the value of endophenotypes in people as opposed to animals?

Battaglia: Is one of your questions about how we affect diagnosis by what we are finding with endophenotypes?

Poulton: Our current diagnostic groupings tend to mix apples and oranges. There is enormous heterogeneity in something we diagnose as 'X'. People have tried to deal with this problem by studying narrower endophenotypes. Maybe our G × E work, or gene discovery work, can be enhanced as a result. While this is a seductive idea, I think it avoids a major issue—the quality or validity of our psychiatric syndromes.

Battaglia: I think yes and no! If at least part of this story is true, then you can back on symptoms and find, for example, that choking and suffocatory symptoms are there and make part of the set of symptoms by which you diagnose panic. Maybe you end up finding that the genes that fall short of significance to be relevant for panic as a whole are significantly relevant for respiratory symptoms in panic. It has to do with refining the mess up of a diagnosis, which is something we can hardly get around. Behavior is complex. But one can imagine a way out of this, if this story as a portion of reality does have power to go back to phenotype and work on it. By the way, the drugs that have allowed to separate panic from generalized anxiety were trycyclics (Klein 1964). They work on two systems: serotonergic and cholinergic. The strongly anticholinergic tricyclic imipramine was the one that led to the dissection of panic from general anxiety. This molecule blocks muscarinic receptors in addition to acting upon serotonergic reuptake.

References

Academy of Medical Sciences 2007 Identifying the environmental causes of disease: how should we decide what to believe and when to take action? London: Academy of Medical Sciences

Battaglia M, Ogliari A 2005 Anxiety and panic: from human studies to animal research and back. Neurosci Biobehav Rev 29:169–179

Battaglia M, Ogliari A, Zanoni A et al 2005 Influence of the serotonin transporter promoter gene and shyness on children's cerebral responses to facial expressions. Arch Gen Psychiatry 62:85–94

Fergusson DM, Horwood LJ, Boden JM 2006 Structure of internalising symptoms in early adulthood. Br J Psychiatry 189:540–546

Gregory AM, Caspi A, Moffitt TE, Koenen K, Eley TC, Poulton R 2007 Juvenile mental health histories of adults with anxiety disorders. Am J Psychiatry 164:301–308

Kaufer D, Frideman A, Seidman S, Soreq H 1998 Acute stress facilitates long-lasting changes in cholinergic gene expression. Nature 393:373–377

Kendler KS, Neale MC, Kessler RC, Heath AC, Eaves LJ 1992 Childhood parental loss and adult psychopathology in women. A twin study perspective. Arch Gen Psychiatry 49:109–116

Klein DF 1964 Delineation of two drug-responsive anxiety syndromes. Psychopharmacology 5:397–408

Rothman KJ, Greenland S 1998 Modern epidemiology, 2nd Edition. Philadephia: Lippincott-Raven

Schneider H, Patil SP, Canisius S et al 2003 Hypercapnic duty cycle is an intermediate physiological phenotype linked to mouse chromosome 5. J Appl Physiol 95:11–19

Tankersley CG, Broman KW 2004 Interactions in hypoxic and hypercapnic breathing are genetically linked to mouse chromosomes 1 and 5. J Appl Physiol 97:77–84

4. Use of monozygotic twins to investigate the relationship between 5HTTLPR genotype, depression and stressful life events: an application of Item Response Theory

Naomi R. Wray*, William L. Coventry*†, Michael R. James*, Grant W. Montgomery*, Lindon J. Eaves‡ and Nicholas G. Martin*

*Genetic and Molecular Epidemiology Laboratories, Queensland Institute of Medical Research, Brisbane, 4006, Australia, †School of Behavioural, Cognitive and Social Sciences, University of New England, Armidale, 2350, Australia and ‡Virginia Institute for Psychiatric and Behavioral Genetics, Richmond, VA 23298, USA

Abstract. We examine the interaction between stressful life events (SLE) and genotypes for the length polymorphism of the serotonin receptor gene (5HTTLPR) on risk of depression. We hypothesize that if the interaction is real, monozygotic twin pairs (MZT) homozygous for the short allele (SS) will have a greater within pair variance in depression measures than MZT homozygous for the long allele (LL), as a reflection of their increased sensitivity to unknown environmental risk factors. Telephone interviews were used to assess symptoms of depression and suicidality on 824 MZT. Rather than using the interview items to calculate sum scores or allocate diagnostic classes we use Item Response Theory to model the contribution of each item to each individual's underlying liability to depression. SLE were also measured on the MZT assessed by mailed questionnaire on average 3.8 years previously, and these were used in follow-up analyses. We find no evidence for significant differences in within pair variance between 5HTTLPR genotypic classes and so can provide no support for interaction between these genotypes and the environment. The use of MZT provides a novel framework for examining genotype × environment interaction in the absence of measures on SLE.

2008 Genetic effects on environmental vulnerability to disease. Wiley, Chichester (Novartis Foundation Symposium) p 48–67

Major depression (MD) is a common psychiatric disorder projected to become the second leading cause of disability worldwide by 2020 (Murray & Lopez 1996). There is strong evidence for a genetic component of liability to MD with a

48

meta-analysis estimate of heritability of 37% (Sullivan et al 2000). Despite this, few genes for depression have been discovered. One of the most studied polymorphisms is the length polymorphism repeat (LPR) in the promotor region of the serotonin transporter gene (5HTT renamed SLC6A4). The 5HTTLPR polymorphism comprises a 43 base pair (Nakamura et al 2000, Hu et al 2005, 2006, Kraft et al 2005, Wendland et al 2006) insertion or deletion (long, 'L', or short, 'S', alleles respectively). The S allele reduces transcriptional efficiency resulting in decreased SLC6A4 expression and 5HT uptake in lymphoblasts (Lesch et al 1996). Many studies have explored the association between 5HTTLPR and depression. Large-sample studies (Willis-Owen et al 2005) and meta-analyses (Anguelova et al 2003, Levinson 2005) have each concluded there was no association between depression and 5HTTLPR, although the latest meta-analysis (Lopez-Leon et al 2007) reported 5HTTLPR to be one of only five variants to show consistent evidence for association with MD. One explanation for these conflicting results is the biologically appealing hypothesis of an interaction between genotype and environment. Caspi et al (2003) reported that individuals who experienced stressful life events (SLE) had an increased risk of depression with each additional S allele, but for individuals who had never experienced SLE, the S allele was not associated with depression. Large scale studies that measure both SLE and depression and take blood samples for genotyping are costly in both time and money and their lack of availability has limited the opportunities for replication studies of this reported interaction between SLE and depression. Nonetheless, 12 replication studies of these results, reported to date, have yielded conflicting results (reviewed by Coventry 2007). Monozygotic twin pairs (MZT) provide an experimental design that allows investigation of environmental variance between genetically identical individuals within 5HTTLPR genotype class (Jinks & Fulker 1970, Eaves & Sullivan 2001).

Statistical interaction does not always equate to biological interaction and can change depending on the scale of measurement (Rothman et al 1980). A continuous liability distribution is widely accepted to underlie the dichotomous measurement of disease including major depression (Eaves et al 1987). In a simulation study, Eaves (2006) demonstrated that genotype × environment interaction (G × E) could be detected with a dichotomous disease status even when no interaction was present in the underlying distribution of liability to disease, thereby questioning the interpretation of the SLE × 5HTTLPR results. Item Response Theory (IRT) in combination with Markov Chain Monte Carlo (MCMC) estimation is considered to provide a flexible and efficient framework for modeling the underlying continuous liability to disease for behavioral phenotypes based on responses to multiple items in an interview framework (Eaves et al 2005). In a simulation study, estimation of heritability was found to reflect more accurately the heritability of the underlying continuous variable when IRT was used rather than analysis of sum scores of the individual items (van den Berg et al 2007).

We hypothesize, that if the interaction between SLE and 5HTTLPR genotype is real, MZT of genotypes SS and SL will have a greater within pair variance in depression measures than MZT of genotype LL, as a reflection of their increased sensitivity to unknown environmental risk factors experienced by individuals. Here, we investigate the relationship between 5HTTLPR genotype, SLE and depression using a cohort of 824 monozygotic twins. Telephone interviews were used to assess symptoms of depression. Rather than using the interview items to calculate sum scores or allocate diagnostic classes we have used IRT, modeling the contribution of each item to each individual's underlying liability to depression.

Materials and methods

Samples

The participants are 824 monozygotic twin pairs from the Australian NHMRC Twin Register (ATR) who are of predominantly North European ancestry and are part of a study described elsewhere (Bierut et al 1999). During the period 1988–1990 study participants were mailed an extensive Health and Lifestyle Questionnaire (HLQ) containing 40 items addressing SLE in three inventories (personal, network and social problems) which were adapted from the List of Threatening Experiences (LTE) (Brugha et al 1985). The 12-item inventory of personal life events (PLE) probed events experienced directly by the participant the previous 12 months: divorce; marital separation; broken engagement or steady relationship; separation from other loved one or close friend; serious illness or injury; serious accident (not involving personal injury); being burgled or robbed; laid off or sacked from job; other serious difficulties at work; major financial problems; legal troubles or involvement with police; and living in unpleasant surroundings. The 21-item network life events (NLE) inventory investigated events experienced by someone in the participant's social network within the previous 12 months, a spouse, child, mother or father, twin, sibling, relative, or someone close had died, suffered a serious illness/injury, or suffered a serious personal crisis. The 7-item social problem inventory included items which addressed serious problems in relationships with a spouse, other family member, close friend, neighbor, someone living with them (e.g. child or elderly parent), their twin, or a workmate or co-worker, during the previous 12 months. Based on results of a preliminary factor analysis, the social problem events were included together with the PLE. The PLE variables used in the analysis were the number of events experienced, truncated to a maximum of 8 events. PLE scores were missing for 5.9% of individuals and were replaced by mean values. NLE were not used in this analysis as preliminary analyses (not shown) found them to have minor impact on risk of depression compared to PLE.

Over the period 1992–2000 participants were interviewed by telephone using the SSAGA-OZ interview instrument, a modified version of the SSAGA (Semi-Structured Assessment for the Genetics of Alcoholism), a comprehensive psychiatric interview designed to assess the physical, psychological and social manifestations of alcoholism and psychiatric disorders in adults (Bucholz et al 1994). The mean interval between the HLQ and SSAGA-OZ interviews was 3.8 years. The SSAGA-OZ telephone interview instrument included two gateway items probing depression. Those answering 'yes' to either of the gateway items (39%) were presented with an additional seven binary items. The specific wording for these gateway items was [1] 'Have you ever had a period of at least two weeks when you were feeling depressed or down most of the day nearly every day?' and [2] 'Have you ever had a period of at least two weeks when you were a lot less interested in most things or unable to enjoy the things you used to enjoy?' The interview instrument also included 6 items (two gateway items each followed by an additional two follow-up items) used to assess lifetime history of suicidality. Abbreviated statements of each item are listed in the key to Plate 5.

Genotyping

Genomic DNA was extracted using standard protocols (Miller et al 1988) from peripheral venous blood samples. Zygosity was determined from self-report questions about similarity and the extent to which the co-twins were confused with each other. Inconsistency of responses resulted in follow-up clarification by telephone and if doubt remained, we asked them to mail in photos at different life stages. This method has demonstrated over 95% agreement with extensive blood

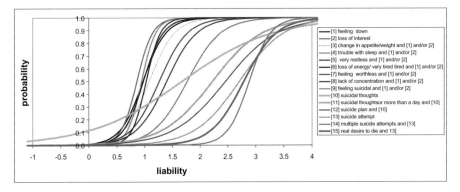

PLATE 5. Item response curves for each item using estimated values of *a* and *b* of equation [1]. The x-axis represents the normally distributed trait, liability to depression and the y-axis is the probability of endorsement of an item. A full-color version of this figure is available in the color plate section of this book.

sampling diagnoses (Martin & Martin 1975, Ooki et al 1990). Zygosity status is updated based on ongoing genotyping studies conducted in our laboratory. Each individual was genotyped for the 5HTTLPR using three different assays to reduce genotyping errors experienced for the original PCR assay because of the very high GC content and the long length of the PCR products which results in bias towards S allele identification and heterozygote drop-out. Full details are given elsewhere (manuscript in preparation). The number of twin pairs in each genotype class was 148, 410 and 266 for SS, SL and LL respectively; these frequencies are in Hardy-Weinberg equilibrium and are representative of our total population sample that included dizygotic twin pairs and siblings.

The minor allele of a single-nucleotide polymorphism (SNP), rs25531 that lies within the L allele of 5HTTLPR has been reported (Hu et al 2006, Wendland et al 2006) to make the L allele functionally equivalent to the S allele because of changes to the AP2 transcription factor binding site altered by this SNP. We typed this SNP in our sample, but found that the reclassification of genotypes made little difference to our results and so is not considered further here.

Statistical analyses

Rather than impose somewhat arbitrary weights to the questionnaire item responses to generate di- or polychotomous diagnosis variables we used IRT in which responses to each item are used to model an underlying (or latent) liability variable for each individual (Lord & Novick 1968, Eaves et al 1987). IRT models were analyzed in the BUGS (Bayesian Inference Using the Gibbs Sampler) program, winBUGS version 1.4.1 (WinBUGS 2004).

Depression and suicidality items were scored as zero if the items were not asked because the respondent did not pass the gateway screening items. Assuming that questions not asked would have been answered as 'no' (=zero) seems reasonable for the suicidality items [11] & [12] and [14] & [15]. It is less clear if respondents would have answered 'no' to items [3] to [9] had they been asked the questions, even though they answered no to both gateway items. We could have included these items as missing (as missingness can be interpolated in the item response modeling) but this would have implied that responses from individuals who answered 'no' to both gateway items can be interpolated from those that answered 'no' to one gateway item. Instead, we chose to accept a more restricted definition of underlying liability to depression, restricted to ability to articulate either feeling down or loss of interest. Items asked but not answered had responses included as missing. A maximum of 2.5% of responses were missing for each item.

Response to item j by individual j (r_{ij}) is assumed to be distributed $r_{ij} \sim$ Bernouilli(p_{ij}) and

$$\text{Logit}(p_{ij}) = b_i(y_{ij} - a_i) \qquad [1]$$

where y_{ij} is the normally distributed underlying latent variable for individual j for item i

$$y_{ij} = b_{\text{sex}}*sex_j + b_{\text{PLEk}}PLE_j + Fam_j + e_j$$

where individual j has sex sex_j (0 = male, 1 = female), genotype class k_j (k = ss, sl or ll), number of personal SLE PLE_j; b_{sex} is the fixed effect of sex (female deviation from male mean) and b_{PLEk} is the regression coefficient for PLE specific to the kth genotype class. Fam_j is the effect of the family and e_j is the individual error term, both of which are modeled within genotype class. Where $Fam_j \sim N(\mu_{kj}, \sigma^2_{kj} r_{kj})$ and $e_j \sim N(0,(1 - r_{kj}) \sigma^2_{kj})$ with constraints imposed so that $\mu_{sl} = 0$, $\sigma^2_{sl} = 1$. Models which ignored stressful life events were considered initially ($b_{\text{PLE}k} = 0$). After a burn-in phase of 1000 iterations and checking that convergence had been achieved, the characterization of the posterior distribution for the model parameters was based on 1000 iterations from two independent Markov chains. The WinBUGS code for this model is included in the Appendix.

To investigate empirically the power of the item response data within the Bayesian IRT framework, we repeated analyses for randomly chosen data sets comprising half (412 MZT pairs) and quarter (206 MZT pairs) with the same distribution of genotype classes. We extrapolated the relationship between sample size and standard errors of estimates to suggest the sample size required to have standard errors sufficiently small to make the magnitude of differences observed significant.

Results

A description of the MZT in terms of sex and age at participation in the HLQ are presented by genotype class in Table 1. We first considered the model excluding

TABLE 1 Description of the MZT by 5HTTLPR class and sex

5HTTLPR genotype	N = Number MZ twin pairs Age = Mean age in years at HLQ survey	Male	Female	Total
SS	N	40	108	148
	Age	43.1	40.5	41.2
SL	N	104	306	820
	Age	40.0	40.0	40.0
LL	N	81	185	266
	Age	39.0	40.8	40.3
Total	N	225	599	824
	Age	40.2	40.4	40.3

stressful life events under the hypothesis that MZ twins pairs with increasing numbers of S alleles at the 5HTTLPR will have increased within pair variance resulting from increased sensitivity to unique, unknown environmental risk factors. The mean value of the parameter estimates and their SD, median and 95% confidence intervals over iterations are given in Table 2. The distribution of liability for the SL group was constrained to have mean 0 and variance 1, so results are expressed as liability to depression in standard deviation units of the SL group. There were no significant differences in means between genotype classes (SS: 0.03 ± 0.09, SL: 0; LL −0.10 ± 0.08) and no difference in within pair variance (SS: 0.58 ± 0.12, SL: 0.52 ± 0.05; LL 0.61 ± 0.12) or between pair variance (SS: 0.31 ± 0.12, SL: 0.48 ± 0.05; LL 0.48 ± 0.12) with increasing numbers of S alleles. Indeed

TABLE 2 Parameter estimates for covariates and variances estimated using IRT

	Model excluding PLE					Model including PLE				
	Mean	SD	2.5%tile	Median	97.5%tile	Mean	SD	2.5%tile	Median	97.5%tile
SEX	0.11	0.08	−0.04	0.12	0.24	0.13	0.05	0.02	0.13	0.22
Mean										
SS	0.03	0.09	−0.15	0.03	0.21	0.12	0.10	−0.09	0.12	0.30
SL[a]	0.00					0.00				
LL	−0.10	0.08	−0.26	−0.10	0.06	−0.11	0.10	−0.34	−0.10	0.08
PLE										
SS						0.19	0.04	0.11	0.19	0.28
SL						0.24	0.02	0.20	0.24	0.28
LL						0.24	0.04	0.17	0.24	0.31
MZ correlation										
SS	0.34	0.10	0.14	0.34	0.53	0.31	0.10	0.11	0.31	0.49
SL	0.48	0.05	0.37	0.48	0.58	0.36	0.06	0.24	0.36	0.48
LL	0.44	0.08	0.29	0.44	0.59	0.36	0.07	0.23	0.36	0.50
Total variance										
SS	0.89	0.16	0.61	0.88	1.27	0.93	0.14	0.67	0.92	1.22
SL[a]	1.00					1.00				
LL	1.09	0.16	0.83	1.07	1.45	1.17	0.16	0.91	1.15	1.55
Between pair variance[b]										
SS	0.31	0.12	0.11	0.30	0.58	0.29	0.11	0.10	0.28	0.51
SL	0.48	0.05	0.37	0.48	0.58	0.36	0.06	0.24	0.36	0.48
LL	0.48	0.12	0.30	0.47	0.75	0.43	0.11	0.25	0.41	0.67
Within pair variance[c]										
SS	0.58	0.12	0.39	0.57	0.85	0.64	0.13	0.43	0.63	0.95
SL	0.52	0.05	0.42	0.52	0.63	0.64	0.06	0.52	0.64	0.76
LL	0.61	0.12	0.41	0.59	0.86	0.75	0.13	0.53	0.74	1.03

[a] Constrained to these values for identification of the model.
[b] Between pair variance = MZ correlation* Total variance.
[c] Within pair variance = Total variance—Between pair variance.

any trends in variances between genotype classes were in the opposite direction to that projected by our prior hypothesis. The effect of sex also failed to reach significance.

Next we considered the model which included known PLE. The regression on PLE was significantly different from zero for each genotype class (SS: 0.19 ± 0.04, SL: 0.24 ± 0.02; LL: 0.24 ± 0.04) but the regression coefficients did not differ significantly from each other, although once again the trend is in the opposite direction to that predicted by the results of Caspi et al (2003). The trend in mean liability to depression between genotype classes (SS: 0.12 ± 0.10, SL: 0; LL: -0.11 ± 0.10) was not significant. The within-pair, between-pair and total variance all showed a trend for SS $\leq$ SL $<$ LL, the opposite direction to our prior hypothesis, but non-significant. To get an empirical handle on the sample size required for the observed differences in variances to be statistically significant, we compared the magnitude of the standard error of estimated parameters for data sets of one-half and one-quarter of the size. We confirmed that, for our item response data, the usual relationship between standard error and sample size (proportional to the ratios of $\sqrt{N}$, where N is the sample size). For the observed difference in total variance of SS and LL, 0.93 and 1.17 to be significant, the standard errors of the estimates need to be at most 0.06, a 2.5-fold reduction, implying a required sample size of $2.5 * 2.5 = 6.25$ times our sample size or 5150 MZT.

Plate 1 presents graphically the individual estimates of the items a_i and b_i for each item i (using the model which includes known PLE). The x-axis represents the normally distributed trait, liability to depression and the y-axis is the probability of endorsement of an item. Estimates of $a_i * b_i$ are reflected in the thresholds at which the response curve has probability >0 (or 'difficulties of the item' or specificity), and the estimate b_i reflects the steepness of the response curve (or sensitivity). The coding of non-asked questions as zero ensures that the curves for either items [1] or [2] always precede items [3] to [9]. The specific wording for item [9] was 'Were you frequently thinking about death, or taking your life, or wishing you were dead?' which was only asked to those who answered 'yes' to one of the two gateway questions [1] and [2]. This is compared to item [10] asked to all participants and worded as 'Have you ever thought of taking your own life?'. Comparing the shapes of the item response curves for these items we see that item [9] is both more sensitive (non-zero probability at higher liability) and more specific (steeper curve) than item [10]. This is partly a reflection of the subtle nuances in the wording of the questions; item [9] is worded more strongly than item [10] so a lower endorsement would be expected. However, the lower endorsement of item [9] may also partially reflect that it is a conditional item. Other than gateway item [10], all the suicidality items show higher specificity to depression (are further to the right) than the general depression items [1]–[9]. The Cronbach's α of the 15 items was 0.91.

Discussion

The paradigm of genotype × environment interaction as presented by the interaction between 5HTTLPR genotype and SLE and its effect on depression is an appealing one. Here we have attempted to use MZ twin pairs to test the hypothesis that genetically identical individuals will show within pair variance dependent on their 5HTTLPR genotype class SS > SL > LL. This hypothesis can be tested in MZT without the need for direct measurements of SLE (even though in our study measures of SLE were available and we used these in follow-up analysis). We found no statistical difference in the within-pair variances by genotype class. Moreover, any trend we observe is in the opposite direction to that predicted by our prior hypothesis, with higher within-pair variance for the LL genotype class compared to the SS class. The same trend is seen for the between pair variance and the total variance, so that the correlation between pairs is lowest for the SS class. The total variance is about 25% greater for the LL genotype compared to the SS genotype class. Given the magnitude of the standard errors, we estimate that a sample size of 5150 MZ twins is required for this difference to be statistically significant.

The validity of our study sample has been demonstrated with heritability of liability to depression estimated to be 36% using DZ twins and siblings (Middeldorp et al 2005b) collected as part of the same study as the MZ siblings used here. Our prior hypothesis assumed that there are environmental risks uniquely experienced by each MZ twin. The correlation between MZ pairs represents the influence of shared genetic background and shared environmental risks. We had measures on SLE experienced by each individual; the relationship between SLE and liability to depression was significant at 0.24 ± 0.02 standard deviation units per SLE for genotype class SL, with no significant differences between genotype classes. On average, SLE were recorded 3.8 years before the depression questionnaire. The relationship between SLE and depression reflects that the depression instrument was probing lifetime depression, but may also suggest that experiencing or reporting SLE may be part of the trait rather than state of depression (Kendler et al 1993, Middeldorp et al 2005a). The reported SLE were also shared, in part, between MZ twins, with the correlations between MZ being lower when known SLE were included in the model.

We have demonstrated the application of IRT to detection of genotype × environment interaction in modeling an underlying latent variable represented by questionnaire responses. IRT models are flexible and easy to apply in the WinBUGS framework, although some thought is required to ensure that items included are representative of the latent variable that is being modeled and to determine the circumstances in which it is appropriate to include responses to non-asked questions as 'missing' or 'known to be negative', both scenarios make assumptions and the individual circumstances will dictate which assumptions are most valid.

In conclusion, we have used MZT and IRT to investigate the interaction 5HTTLPR genotype class and liability to depression. We find no evidence for higher within pair variance for the SS genotype class and therefore can provide no support for the reported interaction between S alleles and unique environment nor measured SLE and liability to depression. However, we acknowledge some caveats of our study that may introduce differences with the results observed by Caspi et al (2003). Our participants had a mean age of 40 years when measured for SLE, with symptoms of lifetime depression and suicidality measured on average 3.8 years later. In contrast, the participants in the study of Caspi et al (2003) were a birth cohort aged 26 years who completed questionnaires probing SLE over the previous 5 years with depression being assessed for the previous year. Only 30% of their sample reported no SLE and 15% reported four or more SLE. In our study the corresponding percentages are 44% and 5%, for SLE which are qualitatively similar. Therefore, we cannot rule out that the interaction between SLE and 5HTTLPR genotype may be age dependent.

Acknowledgments

This research was support by grants to NGM and GWM from the National Health and Medical Research Council (NHMRC; 941177, 971232, 3399450 and 443011) and to Andrew Heath from the US Public Health Service (AA07535, AA07728 & AA10249). Model development and programming were supported by DHSS grant MH068521 to LJE. WLC was supported by (a) an Australian Postgraduate Award from the University of New England, and (b) a postdoctoral fellowship from the University of New England. We would like to thank Leanne Ryan and Troy Dumenil for their careful sample preparation and genotyping. Most of all, we would like to thank the twins for their willing participation in our studies.

References

Anguelova M, Benkelfat C, Turecki G 2003 A systematic review of association studies investigating genes coding for serotonin receptors and the serotonin transporter, 1: affective disorders. Mol Psychiatry 8:574–591

Bierut LJ, Heath AC, Bucholz KK et al 1999 Major depressive disorder in a community-based twin sample: are there different genetic and environmental contributions for men and women? Arch Gen Psychiatry 56:557–563

Brugha T, Bebbington P, Tennant C, Hurry J 1985 The List of Threatening Experiences: a subset of 12 life event categories with considerable long-term contextual threat. Psychol Med 15:189–194

Bucholz KK, Cloninger CR, Dinwiddie DH et al 1994 A new, semi-structured psychiatric interview for use in genetic linkage studies: a report of the reliability of the SSAGA. J Stud Alcohol 55:149–158

Caspi A, Sugden K, Moffitt TE et al 2003 Influence of life stress on depression: moderation by a polymorphism in the 5-HTT gene. Science 301:386–389

Coventry WL 2007 Perceived social support: Its genetic and environmental etiology and association with depression. Armidale: University of New England

Eaves LJ 2006 Genotype by environment interaction in psychopathology: fact or fiction? Twin Res Hum Genet 9:1–9

Eaves LJ, Sullivan P 2001 Genotype-environment interaction in transmission disequilibrium tests. Adv Genet 42:223–240

Eaves LJ, Martin NG, Heath AC, Kendler KS 1987 Testing genetic models for multiple symptoms: an application to the genetic analysis of liability to depression. Behav Genet 17:331–341

Eaves L, Erkanli A, Silberg J et al 2005 Application of Bayesian inference using Gibbs sampling to item-response theory modeling of multi-symptom genetic data. Behav Genet 35:765–80

Hu XZ, Oroszi G, Chun J et al 2005 An expanded evaluation of the relationship of four alleles to the level of response to alcohol and the alcoholism risk. Alcohol Clin Exp Res 29:8–16

Hu XZ, Lipsky RH, Zhu G et al 2006 Serotonin transporter promoter gain-of-function genotypes are linked to obsessive-compulsive disorder. Am J Hum Genet 78:815–826

Jinks JL, Fulker DW 1970 Comparison of the biometrical genetical, MAVA and classical approaches to the analysis of human behavior. Psychol Med 73:311–349

Kendler KS, Neale M, Kessler R, Heath A, Eaves L 1993 A twin study of recent life events and difficulties. Arch Gen Psychiatry 50:789–796

Kraft JB, Slager SL, McGrath PJ, Hamilton SP 2005 Sequence analysis of the serotonin transporter and associations with antidepressant response. Biol Psychiatry 58:374–381

Lesch KP, Bengel K, Heils A et al 1996 Association of anxiety-related traits with a polymorphism in the serotonin transporter gene regulatory region. Science 274:1527–1531

Levinson DF 2005 Meta-analysis in psychiatric genetics. Curr Psychiatry Rep 7:143–151

Lopez-Leon S, Janssens AC, Gonzalez-Zuloeta Ladd AM et al 2007 Meta-analyses of genetic studies on major depressive disorder. Mol Psychiatry, in press

Lord FM, Novick MR 1968 Statistical theories of mental test scores. Addison-Wesley Publishing Company, Inc

Martin NG, Martin PG 1975 The inheritance of scholastic abilities in a sample of twins. I. Ascertainment of the sample and diagnosis of zygosity. Ann Hum Genet 39:213–218

Middeldorp CM, Birley AJ, Cath DC et al 2005a Familial clustering of major depression and anxiety disorders in Australian and Dutch twins and siblings. Twin Res Hum Genet 8:609–615

Middeldorp CM, Cath DC, Vink JM, Boomsma DI 2005b Twin and genetic effects on life events. Twin Res Hum Genet 8:224–231

Miller SA, Dykes DD, Polesky HF 1988 A simple salting out procedure for extracting DNA from human nucleated cells. Nucleic Acids Res 16:1215

Murray CJ, Lopez AD 1996 Evidence-based health policy—lessons from the Global Burden of Disease Study. Science 274:740–743

Nakamura M, Ueno S, Sano A, Tanabe H 2000 The human serotonin transporter gene linked polymorphism (5-HTTLPR) shows ten novel allelic variants. Mol Psychiatry 5:32–38

Ooki S, Yamada K, Asaka A, Hayakawa K 1990 Zygosity diagnosis of twins by questionnaire. Acta Genet Med Gemellol (Roma) 39:109–115

Rothman KJ, Greenland S, Walker AM 1980 Concepts of interaction. Am J Epidemiol 112:467–470

Sullivan PF, Neale MC, Kendler KS 2000 Genetic epidemiology of major depression: review and meta-analysis. Am J Psychiatry 157:1552–1562

Van Den Berg SM, Glas CAW, Boomsma DI 2007 Variance decomposition using an IRT measurement model. Behav Genet 37:604–616

Wendland JR, Martin BJ, Kruse MR, Lesch K-P, Murphy DL 2006 Simultaneously genotyping of four functional loci of human SLC6A4, with a reappraisal of 5-HTTLPR and rs25531. Mol Psychiatry 11:1–3

Willis-Owen SAG, Turri MG, Munafo MR et al 2005 The serotonin transporter length poly-
morphism, neuroticism, and depression: a comprehensive assessment of association. Biol
Psychiatry 58:451–456
WinBUGS 2004 Version 1.4.1; Imperial College and MRC, UK

DISCUSSION

Uher: We have used the item response modeling in a pharmacogenetics study to
improve measures of depression and it has substantially clarified our results (Uher
et al 2008). The item response modeling does two different things: it is a threshold
model and also factor analysis in one. Firstly, it weights items depending on how
much they load on one underlying dimension, and this is reflected in the discrimination parameters. The items that don't fit the one dimension are loaded less.
Secondly, it weights the item response options according to how extreme values
of the underlying dimensions they are likely to indicate, and this is reflected in the
threshold parameters. The two aspects of each item are integrated to make the
best estimate of the true score on a latent underlying dimension.

Martin: I forgot to make this point. The sensitivity is really just the factor
loading.

Uher: In our data, this makes up most of the difference. The other is the ranking
of thresholds. This is useful, because if you have many items that measure the same
thing, then summing them up doesn't make sense and different scales get biased
towards the symptoms, which they measure by more items. In our data we found
that depression scales were much better described by three factors that are reasonably non-overlapping: the observed mood, the cognitive symptoms, and vegeta-
tive/somatic symptoms. Your suicidal thoughts measure would probably go more
with the cognitive ones, which is why it didn't load so highly. In my experience it
is a cue to partitioning into dimensions.

My other comment is about the time lag between life events and depression.
What you found is fairly typical, and it is also in the findings of George Brown
and Tirrill Harris, that there is a strong correlation between the life events preceding the onset of depression by three months or less, but there is also a weaker but
significant correlation with life events that are remote in time. George Brown's
interpretation of that is that it is the influence of early experiences. In their work
they also addressed the independence of life events. It would be interesting to
know whether these are the life events that are likely to be contributed to by the
subject themselves.

Heath: Lindon Eaves has a history of making important innovations in this field.
He has convinced me that for this type of problem he is right that Bayesian
methods have considerable potential to allow more rigorous testing of hypotheses
about genotype × environment (G × E) interaction. This work forces us to think

about the properties of our measurement scales. If we work in areas such as psychiatric genetics, where in the end we have a number of symptoms and are trying to draw inferences, the more we understand about how our measurement scale and how it can cause us to make incorrect inference, the more we can be confident about reaching correct conclusions. It comes back to this idea of being sensitive to the assumptions we are making and trying to test them. There is nothing inherent in the IRT model that says it has to be unidimensional.

Uher: Unidimensionality is an assumption for fitting an IRT model.

Martin: You can specify multiple factors. I have just shown a single factor but it could easily be parameterized for several factors.

Heath: In the Bayesian framework you could easily make this a tridimensional model, with sibships of various sizes. If you are working within a frequentist framework that estimation problem rapidly becomes intractable. Nick Martin, you illustrated some of the nice summary plots that can come from a Bayesian simulation-based analysis. But there are many other things you can get to allow you to look more critically at the assumptions you are making. For example, am I doing OK assuming a normally distributed liability to depression? Or have I really got something that is approximated by a mixture of normals? This is easy to plug into a Bayesian framework, but much more difficult to do in a frequentist framework. If I am interested in using a quantitative trait, then I am saying that I have got good discrimination between people scoring on different points on my scale. It is an easy step to say, what are my 95% CIs on how I am ranking people? This can be a depressing discovery, looking at DSM-IV nicotine dependence and finding that I do OK at the top and bottom ends, but have a mish-mash in the middle. We gain power if our G × E effects are acting on a quantitative scale. But it is helpful to look at how that quantitative scale we are trying to create is behaving, because then we can start thinking about how to improve this performance.

Martin: One of the points I didn't mention was a problem we recently discovered: the results of this analysis are very sensitive to how certain items are treated. Our interview has a couple of gateway items, and you are only asked subsequent questions depending on the results of the gateway question. The results we get out depend critically on how we treat these other seven items: whether we treat them as missing or treat them as zero, given that they didn't pass the two gateway items. We were quite shocked about what a difference this made. If you treat them as missing, the item difficulties all clump up together, right around the discrimination point. If you treat them as zero you get a much bigger spread of the item difficulties, and change in the item sensitivities.

Heath: Lindon Eaves has some wonderful insights. There is a nice example of where his original script is mis-specifying the missing-ness of the data. Nick and I arrived independently at this recognition. This was also true of the original latent class analysis paper.

Uber: The finding of Nick Martin and colleagues is not so surprising. If they replace the items that are not asked with zeros, the variability of those items is artificially decreased and subsequently these items contribute little to the model. If you treat them as missing data, which is what they are, then the variability of the non-missing values reflects their true variability and makes these items more important part of the model. The problem remains though that while these items are missing, they are not missing at random.

Heath: I don't think either Lindon's or Nick's approach is the correct way to handle the problem. I think it is a question of how you write the likelihood. The bigger picture is that these methods are potentially very powerful, but their implementation is posing statistical challenges that biostatisticians in large cancer research groups will be comfortable with, but the guys in the behavior genetics field are struggling to catch up with.

Martin: We need to persuade psychiatrists to ask all questions to all people.

Poulton: Lindon Eaves and others who ask us to be self-critical about how we measure are doing us a favor. For example, in our work we recognize that there is error in any measure we apply. As a general approach we try to measure in multiple different ways. We will ask individuals about their symptoms; we will ask other people about our participants' symptoms; we will use official records where these exist, and so forth. We also present the data in different ways; we will cut it—as often required by diagnoses—as well as present it continuously. We are looking for consistency. At the end of the day, the acid test is to plot the data and see what they look like. Having done all these things, we feel we are on strong footing. This is one way to address concerns about 'pathologies of scaling'. A more specific methodological point relates to how we measured our dependent and independent variables. You mentioned that we measured proximal stresses. Yes, we did—but we measured this over a five year period. We didn't ask people to fill in a questionnaire, but sat them down with a quite detailed life history calendar, in which they get to report on salient events in their life, not just the adverse sort, on a month by month basis. There are personal flags in their calendars to do with factors such as birthdays which recent research from cognitive neuroscience suggests enhance accuracy of recall. You made the point about how outcome is measured, and issues relating to gate questions. Was your depression measure a lifetime measure?

Martin: We had two measures. One was from five years prior to five years after reporting a life event—we call this 'lifetime'. The other was during and after the reporting of life events.

Poulton: Either way, you are asking people to think back over a decent period of time, and there is some real decay in accuracy of reporting over time, particularly about internal states. With that caveat on the table, I have a question. I think IRT has a lot to offer. Given that we have just published a paper using IQ as our outcome (Caspi et al 2007), how does one apply item response theory to normally distributed IQ measures?

Martin: In fact, IRT was developed in the context of IQ. This is why people used the term 'difficulty' for the position along that axis. In fact, one could just apply it to the individual items of an IQ test. In no way was I trying to attack your original finding. I ignored the shortcomings of our data, and they differ from yours in some very important respects. The point was to illustrate the method as being potentially useful.

Poulton: I took your point as a general one related to ways of improving scale measurement.

Martin: I'm here to plug IRT as a really useful tool for tackling these sorts of problems.

Poulton: Yes, we've used it in a different context, to do with personality assessment.

Rutter: Nick Martin, in your paper you outlined problems and you presented solutions. IRT clearly constitutes a useful technique, and we should accept that and not get too bogged down on the fact that it's not perfect. I liked the way that you introduced discordant MZ pairs. It constitutes one of the many types of natural experiment that can be used to test for environmental mediation (Rutter 2007). One comment I would make is that, although MZ pairs are usually described as having no genetic differences, that is not entirely correct. They may differ in gene expression and in various other ways. Nevertheless, it is a powerful tool and its integration with other approaches is useful.

The issue of using life events over time raises a different issue. You quite rightly bring out the fact that life events involve genetic influences and they may work in somewhat different ways if you are looking at accumulated life events over time, which has a trait-like quality, than if you are looking at an acute event having a provoking effect. The sort of approach that George Brown has advocated focuses on its role in precipitating the onset of a disorder. But this needs to be thought about in other ways, too. I don't know whether or not this affects what is found with a gene–environment interaction. This comes back to the kindling notion where the effect of environmental experiences is usually seen as diminishing over time. There have been suggestions that it may work in the opposite way (see Monroe & Harkness 2005): that is, if you have an effect of life events that is leading to the onset of disorder, the fact that you don't see it later is because you are getting effects from lesser life events that are below the threshold of what you are measuring. This gets us involved in complicated side issues. No one has a perfect answer, and we need to recognize the problems and try to find imaginative ways of dealing with them.

Let me focus in a rather mischievous way on an interesting difference between the way you put things in your abstract for the program, and the way you put them in your paper. In the paper you talk about the interesting finding of the fact that the Short-Long difference is influenced by another polymorphism (see Wendland et al 2006) as affecting the 'veracity' of earlier research. There are four points that need to be recognized here. First, this is an interesting finding in its own right. It means that in considering genetic effects, we need to be thinking about them in more complicated ways than we have been used to doing. We are now able to do this better because of the advances in technology. Second, in so far that this is having an effect, it means that the original claim of Caspi, Richie and others is an underestimate, not an overestimate. The veracity criticism seems to be a curious way of expressing that. Third, the frequency of this polymorphism is 6–7%, so the chance of it making much of a difference is quite small. Fourth, as you have shown in your own work, and Zalsman et al (2006) also found, taking this into account actually made no difference. It is an interesting finding that raises all sorts of issues that may have a major effect in other circumstances, but pretty certainly it doesn't have an effect here. Do you agree?

Martin: We sucked it and saw, basically. We didn't see any effect. It is low frequency, so as you say, we would predict this. Perhaps the term 'veracity' was ill-chosen. What I was referring to was how horrified we were when we saw how inaccurate our earlier SL genotyping was, using the standard assays. This was our first report, and we found a few discrepancies—about 30 genotypes that were wrong. The thing that alerted us to this was the fact that we had all these MZ twins in there where we could type both. What I was hinting at with the word 'veracity' was that if we had all these problems, then what about people who didn't have MZ twins or family data to look at the accuracy of this typing method? More than for many assays, we saw real problems with heterozygote drop-out where they were being read as homozygotes. This is a terrible assay, and anyone in this game will acknowledge that. I wonder how reliable some of the early data are.

Rutter: That's right. It is a caution we all need to take account of.

Martinez: I agree with what you are saying. We have a relatively large set of CEPH families which we have tested together with the study subjects in most of our assays.

Martin: Can I advocate using MZ twins. In every assay we do, we throw in a couple of hundred MZ pairs, and there is no quicker reality check!

References

Caspi A, Williams BS, Kim-Cohen J et al 2007 Moderation of breastfeeding effects on cognitive development by genetic variation in fatty acid metabolism. Proc Natl Acad Sci 104: 18860–18865

Monroe SM, Harkness KL 2005 Life stress, the 'kindling' hypothesis and the recurrence of depression: Considerations from a life stress perspective. Psychol Rev 112:417–445

Rutter M 2007 Proceeding from observed correlation to causal inference: the use of natural experiments. Perspect Psychol Sci 2:377–395

Wendland JR, Martin BJ, Kruse MR, Lesch KP, Murphy DL 2006 Simultaneous genotyping of four functional loci of human SLC6A4 with a reappraisal of 5-HTTLPR and rs255531. Mol Psychiatry 11:224–226

Uher R, Farmer A, Maier W et al 2008 Measuring depression: comparison and integration of three scales in the GENDEP study. Psychol Med 38:289–300

Zalsman G, Huang YY, Oquendo MA et al 2006 Association of a triallelic serotonin transporter gene promoter region (5-HTTLPR) polymorphism with stressful life events and severity of depression. Am J Psychiatry 163:1588–1593

Appendix

WinBUGS code for heterogeneity of MZ intrapair variation (G × E)

```
#Data is sorted as N MZT pairs (1st MZ Twin followed by 2nd MZ twin)
# with NSS pairs with 5HTTLPR genotype SS listed first
# followed by NSL pairs with 5HTTLPR genotype SL listed next
# followed by N-NSS-NSL pairs with 5HTTLPR genotype LL
model;
{

# covariates inputted in vector y with dimension N pairs x 2 twin individuals x 2
covariates
  for (i in 1:N){
     for(j in 1:2){
     PLE[i , j]<-y[i, j, 1]
     sex[i , j]<-y[i, j, 2]
  }
}

#prior for covariates
bPLE.SS~dunif(-1,1)
bPLE.SL~dunif(-1,1)
bPLE.LL~dunif(-1,1)
bsex~dunif(-1,1)

# item responses inputted in vector x with dimensions
#N pairs x 2 twin individuals x kitem = 15 0/1 item responses
# model responses with Bernouilli distribution
for(item in 1 : kitem){
     for (i in 1: N){
        for (j in 1:2){
```

```
    x[i , j, item] ~ dbern(p[i , j, item])
        }
      }
    }
```

```
#Simulate latent trait scores for three genotype classes
#Priors on parameters so that Heterozygotes are distributed (N[0,1]);
#MZ correlation
rSL~dunif (0,0.95)
#within pair variance
s2w.SL<-1-rSL
#between pair variance
s2b.SL<-rSL
#mean
muSL<-0


# SS genotype class
# MZ correlation prior
rSS~dunif (0,0.95)
# standard deviation prior
s1.SS~dunif(0.5,1.5)
# Variance and components –total, between, within
s2.SS<-s1.SS*s1.SS
s2b.SS<-rSS*s2.SS
s2w.SS<-(1-rSS)*s2.SS
#mean prior
muSS~dnorm(0,1)


# LL genotype class
# MZ correlation prior
rLL~dunif (0,0.95)
# standard deviation prior
s1.LL~dunif(0.5,1.5)
# Variance and components –total, between, within
s2.LL<-s1.LL*s1.LL
s2b.LL<-rLL*s2.LL
s2w.LL<-(1-rLL)*s2.LL
#mean prior
muLL~dnorm(0,1)
```

```
#WinBUGS works with precision parameters
tb.SS<-1/s2b.SS
tw.SS<-1/s2w.SS
tb.SL<-1/s2b.SL
tw.SL<-1/s2w.SL
tb.LL<-1/s2b.LL
tw.LL<-1/s2w.LL
# Latent trait for homozygotes SS
for (i in 1:NSS){
     mSS[i]~dnorm(muSS,tb.SS)
   for(j in 1:2){
   theta[i , j] ~ dnorm(mSS[i],tw.SS)
   zz[i,j]<- bsex*sex[i , j] + bPLE.SS*PLE[i , j] + theta[i , j]
   }
}


# Latent trait for heterozygotes SL
for (i in NSS + 1:NSS + NSL){
   mSL[i]~dnorm(muSL,tb.SL)
   for(j in 1:2){
   theta[i , j] ~ dnorm(mSL[i],tw.SL)
   zz[i,j]<- bsex*sex[i , j] + bPLE.SL*PLE[i , j] + theta[i , j]
   }
}


# Latent trait for homozygotes LL
for (i in NSS + NSL + 1:N){
 mLL[i]~dnorm(muLL,tb.LL)
   for(j in 1:2){
   theta[i , j] ~ dnorm(mLL[i],tw.LL)
   zz[i,j]<- bsex*sex[i , j] + bPLE.LL*PLE[i , j] + theta[i , j]
   }
}
# Calculate endorsement probabilities
# (Logistic IRT)
  for(item in 1 : kitem) {
    for(i in 1 : N){
        for (j in 1:2){
      logit(p[i , j , item]) <- b[item] * (zz[i , j] - a[item])
      }
      }
```

```
}

# Priors on item parameters
  for(item in 1 : kitem) {
    a[item] ~ dunif(-1,3)
    b[item] ~ dunif(-1,10)
}

# Set any derived parameters that need to be monitored e.g. variance differences
or ratios
}
```

GENERAL DISCUSSION I

Tesson: Now there are lots of papers reporting that even if a particular polymorphism (especially if it is not in the coding sequence) is not associated with a disease in a particular population, there may be other polymorphisms that can be found by doing haplotype analysis, for example, that might be associated with the phenotype. Might these kinds of studies using haplotypes, and also perhaps using other polymorphisms in other genes from the same pathway, be worth doing? This could get us around the problem.

Martin: There are two points there. First, the utility of haplotype analyses. The jury is still out. People have spent a lot of time on these. There are very few examples of where the haplotypes have been more illuminating than the initial single nucleotide polymorphism (SNP).

Tesson: Yes, but at least haplotype analysis might help finding polymorphisms in linkage disequilibrium.

Martin: Because the gel assay is so awful, we have spent a lot of time typing all the other SNPs in that gene and trying to see whether there is sufficient linkage disequilibrium (LD) to just use the SNPs. We can't—it is just not strong enough. The second point you made about looking at other genes in the pathway is a good one. We have become a little cynical about the candidate gene approach. With genome-wide association scans coming on board, why muck around? Let's just go to genome-wide association? Our approach is to get genome-wide association with everyone.

Kotb: I have a remark about our earlier discussion on gene × environment (G × E) and the statistical representation. Being a biologist, I am seeing a bit of generalization of certain approaches which may not be generalizable. The complexity of psychiatric diseases has its challenges but these are quite different from those of other biological problems such as infectious diseases. Different types of diseases have major challenges, but in different ways. How we define 'E' (environment) in an infectious disease, where E is a combination of so many factors including the elaboration of different sets of virulence factors by the pathogen that are expressed at different times during the infection and that interact with each other as well as with a different sets of host defense molecules? Different sets of virulence factors can be expressed depending on the infection site, and similarly the host can express different sets of defense molecules depending on what the microbe is producing. The expression of the host defense molecules can also vary due to host genetic polymorphism and pre-existing immunity etc. To make things even more complicated, the composition of the microbial community can change under the selective pressure of the host to where bacteria with mutations in genes encoding the

68

pathogen's virulence factors that are selected because they are the fittest to survive the hostile host environment. These are very dynamic processes that can vary quite a bit depending on many complex environmental factors. So my question is: how do we define E in these situations? Do we use the same mathematical formulae or approaches used in psychiatric diseases, or do we generate models for E that takes into consideration all the variables that I talked about. Do we need to modify approaches and equations to be take into account the nature of the particular disease that we are study how E × G affect its outcomes. I'd like us to think about this.

Uher: Andrew Heath mentioned earlier that we'd need at least 2000 patients to do whole genome association. I found this surprising, because the power calculations show that we need 1000 people to detect a single-gene interacting effect. The number of individuals needed to detect five genes is already several thousand. For all the genes involved in the causal pathways, the power would be ridiculously small. If you are able to build a framework of how the genes are expected to be interacting along the causal pathway, this helps. Infectious disease may be complicated, but you have the advantage of knowing your causal agent at the molecular level.

Kotb: Each area has its pros and cons in terms of challenges. Do we generalize everything or do we modify our approaches, models and equations to incorporate these challenges into the mathematical models we are trying to use?

Poulton: I think this is a nice point. We are talking about different approaches and designs, all of which have different strengths and weaknesses. At times we fall prey to the tendency to back one over the other. Each has its value and place. For example, when I think of our cohort, part of its value lies in confirming ideas that flow out of genome-wide association studies.

Snieder: With the advent of genome-wide association studies in the last few years, we have seen that they can be quite effective. One good example would be for type 2 diabetes. As soon as we have those replicated candidate genes, we can try to plug them into gene–environment interactions studies. As a genetic epidemiologist with an interest in gene finding, I have found it exciting that we are now finding these genes, which we can use in our G × E studies. This is true for many diseases: psychopathologies, infectious diseases and autoimmune diseases.

Kotb: As long as these associations can be validated biologically, who cares what the *P* value is? One can get highly significant *P* values that may have no biological relevance.

Snieder: As soon as you have these replicated genes you have a specific hypothesis that you can test. You no longer have the problem of multiple testing correction.

Heath: You want to take one or two genes back to the lab, not hundreds!

Tesson: The genes found by genome-wide analysis are not necessarily going to be the same genes that are responders to the environment, and you may have missed them by doing a genome-wide association without taking any account of the environment.

Heath: To give a nice example, I do work on alcohol dependence. Let's hypothesize that there are genes that make you develop problems with alcohol at levels of consumption that are lower than those that most people who have alcohol problems typically drink. They really do make you vulnerable: they put a subgroup of individuals at risk at much lower levels of consumption than normal. The trouble is, when people select cases they tend to select extremes. In their pool they are not getting the people who have sensitivity to the environment of alcohol consumption. Under these conditions, unless I take into account my exposure variable in how I select my cases, I am going to miss this G × E effect, because I am not going to be finding those genes. There are some exciting findings emerging from genome-wide association studies, but there are a lot of investigators working on a broad range of medical conditions who are puzzled about only finding just one or two genes for conditions with multiple genetic risk factors.

Tesson: Look at hypertension!

Martinez: In the cardiovascular field, in recent genome-wide association studies they did something interesting. They tested for those genes that have been reported to be associated with the same phenotypes through candidate gene approaches. In these studies, 70% could not be confirmed. This is a problem because many of these genes had already been replicated. It may be that the genes that are positive hits in genome-wide association studies could be those that interact with universal exposures, thus making them very difficult targets for the study of gene–environment interactions.

Rutter: Nick, were you being mischievous or serious when you said that validity depends on P values? This seems to run counter to what my statistical colleagues tell me. There are many journals that ban P values and say we must present confidence intervals, which give much more information. Others have said that hypotheses aren't created equal (see Academy of Medical Sciences 2007). If you have a result that is highly significant but is completely out of line with biological findings from other research, never mind whether the P value is significant or not, you need to look at it as a whole. If you are making the point that validity is crucially dependent on statistics, then I agree.

Reference

Academy of Medical Sciences 2007 Identifying the environmental causes of disease: how should we decide what to believe and when to take action? London: Academy of Medical Sciences

5. Role of gene–stress interactions in gene-finding studies

Harold Snieder*†‡, Xiaoling Wang†, Vasiliki Lagou*, Brenda W. J. H. Penninx§, Harriëtte Riese*¶ and Catharina A Hartman¶

*Unit of Genetic Epidemiology and Bioinformatics, Department of Epidemiology, University Medical Center Groningen, University of Groningen, The Netherlands, †Georgia Prevention Institute, Department of Pediatrics, Medical College of Georgia, Augusta, Georgia, USA, ‡Twin Research and Genetic Epidemiology Unit, St Thomas' Hospital, Kings College, London School of Medicine, London, UK, §Department of Psychiatry, VU University Medical Center, Amsterdam, The Netherlands and ¶Department of Psychiatry, University Medical Center Groningen, University of Groningen, The Netherlands

Abstract. Identification of genetic variants underlying common complex traits and diseases can be viewed as a three-stage process that jump-started with the sequencing of the human genome. The second phase, characterization of genetic variants in different human populations, has shown major progress in recent years. The increased availability of single nucleotide polymorphisms (SNPs) has already spawned two important developments in genetic association studies. Increasingly, rather than focusing on one or two functional SNPs, candidate gene studies consider all variants within the gene jointly. The second development is that of the whole genome association study. This chapter illustrates two distinct ways in which gene–stress interactions may aid such gene finding studies. We have recently shown for heart rate variability—an index of autonomic dysfunction related to both psychopathology and cardiovascular disease—that exposure to an acute stressful challenge in a standardized lab setting may produce a more heritable endophenotype, facilitating identification of underlying genes. The second example shows how the creation of a cumulative index of chronic stress based on multiple questionnaire- and interview-based measures of stress exposure may be applied in a genome-wide association study of (high) blood pressure to find genes that only come to expression in stressful environments. We conclude that investigation of gene–environment interactions in the context of both gene- and genome-wide association studies may offer important advantages in gene finding efforts for complex traits and diseases.

2008 Genetic effects on environmental vulnerability to disease. Wiley, Chichester (Novartis Foundation Symposium) p 71–86

Finding genes for complex traits and diseases: a three-stage process

Identification of genetic variants underlying common complex traits and diseases can be viewed as a three stage process that jumpstarted with the sequencing of the human genome. Major progress was made in 2005 with the second phase, the characterization of genetic variants in human populations. At the whole-genome level, the International HapMap Consortium (2005) characterized millions of common single nucleotide polymorphisms (SNPs) in individuals of African, European and Asian descent. Other major variation discovery efforts have a candidate gene focus such as the SeattleSNPs project, which has identified all common variants through sequencing several hundred candidate genes for cardiovascular disease in both European and African American samples (Crawford et al 2005). These freely available resources constitute the first steps towards a future where all variants of human genes will be known.

The ultimate success of gene finding efforts for complex traits and diseases will largely depend on the third and final phase in which the connection is made between genetic variation (including gene–gene and gene–environment interactions) and the outcome of interest. This is the (genetic) epidemiology or biobank phase in which it will be crucial to invest heavily in the accurate measurement of lifestyle factors, environmental exposures, detailed characterization of clinical phenotypes and the ability to obtain such information in prospective studies of adequate size.

As opposed to Mendelian disease, progress in identification of genes for complex traits and diseases through positional cloning (i.e. fine mapping of linkage peaks) has been slow. Therefore, gene finding efforts increasingly rely on association approaches facilitated by recent progress in the characterization of genetic variants. The increased availability of SNPs generated in phase two has already spawned two important developments in association studies. Increasingly, rather than focusing on one or two functional SNPs (i.e. direct association), candidate gene studies consider all variants within the gene jointly (i.e. gene-wide) (Neale & Sham 2004). This is best achieved through selection of a minimal set of tagging SNPs effectively capturing all common variation by taking into account patterns of linkage disequilibrium (LD) across the gene. Such an indirect association approach affords gene-based replication and offers a promising solution to the lack of replicability that continues to plague association studies of complex diseases and traits (Neale & Sham 2004). The second development is that of the whole genome association study (Hirschhorn & Daly 2005, Wang et al 2005). Rapid improvements in genotyping technology and reductions in cost have now made it feasible to conduct these studies, using large sets of anonymous SNP markers equally distributed across the genome. Many successes have been reported since the first successful application of such a genome-wide approach in 2005 led to identification of a functional SNP underlying age-related macular degeneration (Klein et al 2005). For example, for type 2 diabetes eight novel genes were identified in the first half of 2007 alone (Frayling 2007).

For each of these two association approaches (gene- and genome-wide) the current chapter will illustrate how investigation of gene–environment interactions may aid gene finding. The first example describes the design of a gene-wide candidate gene association study for heart rate variability (HRV), an index of autonomic dysfunction related to both psychopathology and cardiovascular disease. This study was founded on the observation that exposure to an acute stressful challenge in a standardized lab setting (i.e. gene–stress interaction) produced a more heritable endophenotype, potentially facilitating identification of underlying genes. The second example will show how the creation of a chronic stress index based on multiple questionnaire- and interview-based measures of exposure to stress may be applied in a genome-wide association study of (high) blood pressure to find genes that only come to expression in stressful environments.

A gene-wide candidate gene association study for heart rate variability

Role of challenged endophenotypes for gene finding

Common research designs for association studies include case-control and (population-based) cohort designs. The focus of the former is on dichotomous disease traits, whereas the latter generally focus on quantitative intermediate phenotypes that lie on the etiological path between causal factors and disease outcome. There may be several advantages to studying such intermediate traits (Sing et al 2003) or endophenotypes (Gottesman & Gould 2003) for the genetic dissection of complex diseases. Compared to the final disease outcome, endophenotypes will be closer to the underlying genetic factors and will only be influenced by a subset of genes. These genes will likely have a larger effect on the endophenotype than on the disease, which should facilitate gene finding (Gottesman & Gould 2003, Schork 1997). To distinguish endophenotypes from state-dependent biomarkers of limited reproducibility, Gottesman and Gould (2003) have proposed the following five criteria:

1) The endophenotype is associated with illness in the population;
2) The endophenotype is heritable;
3) The endophenotype is primarily state independent. That is, it manifests in an individual whether or not the illness is active;
4) Within families, endophenotypes and illness cosegregate;
5) An endophenotype identified in probands is found in unaffected family members at a higher rate than in the general population.

In addition to the third criterion they state that the endophenotype may need to be elicited by a challenge such as a glucose tolerance test in relatives of type 2 diabetics (Gould & Gottesman 2006). Many such tests administering a standardized environmental challenge exist in different disease areas. Within the context of genetically informative studies, such tests can be considered experimentally induced gene–environment interactions. In fact, such a challenged endophenotype may be more heritable than its unchallenged counterpart, potentially offering important advantages for gene finding studies.

This principle is elegantly illustrated by Gu et al (2007) who investigated the heritability of blood pressure responses to dietary sodium and potassium intake in 1906 individuals from 658 Chinese pedigrees. The intervention included a 3 day baseline period during which subjects adhered to their usual diet, followed by a 7 day low sodium diet, a 7 day high sodium diet and a 7 day high sodium plus potassium supplement diet. Gu et al (2007) observed that heritabilities of blood pressure under the environmentally controlled sodium and potassium intake were significantly higher than those under the natural diet. Heritabilities of baseline systolic (SBP) and diastolic BP (DBP) were 0.31 and 0.32, respectively. These heritabilities increased significantly to a narrow range of values between 0.49 and 0.52 for both SBP and DBP in all three dietary conditions. Interestingly, the authors showed that these increases in heritability estimates were caused not only by a decrease in unique environmental (or residual) variance, as might have been expected under environmentally controlled circumstances, but also by an equally large increase in additive genetic variance. Although Gu et al (2007) did not elaborate on this, such an increase in genetic variance may have been caused by: (1) a larger effect during the dietary conditions of the same genes that also affect BP at rest; (2) an emergence of new genetic effects on BP specific to the dietary conditions or, (3) a combination of the two. Bivariate models that include both challenged and unchallenged conditions can distinguish between these possibilities and quantify genetic and environmental effects on levels of the challenged and unchallenged phenotypes as well as on the response to the challenge (see Fig. 1). As described in the section below, we recently used such an approach to investigate cardiovascular phenotypes during an acute stress challenge and test for the existence of gene-by-stress interaction within the context of a classic twin study (De Geus et al 2007).

Cardiovascular reactivity to stress: studying gene-by-stress interaction

The cardiovascular 'reactivity hypothesis' put forward by psychophysiologists in the late 1960s and early 1970s posits that cardiovascular reactivity, defined as an exaggerated cardiovascular response to a behavioral or psychological challenge, may play a role as a marker or mechanism in the pathogenesis of essential hypertension and coronary heart disease (Obrist et al 1974). Studies investigating this hypothesis have traditionally analyzed stress reactivity as a change score calculated as the difference between stress and resting levels of the cardiovascular variables

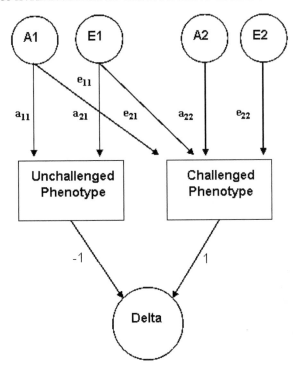

FIG.1.   Path diagram for a bivariate model in a classic twin study. For clarity only one twin is depicted. A1, A2, additive genetic variance components; E1, E2, non-shared environmental variance components; a_{11} through a_{22}, genetic path coefficients (or factor loadings) of which a_{22} represents novel genetic influences on the challenged phenotype; e_{11} through e_{22}, non-shared environmental path coefficients (or factor loadings) of which e_{22} represents novel environmental influences on the challenged phenotype. Delta, difference in levels of challenged and unchallenged phenotype. Subtracting the level of the challenged phenotype from the level of the unchallenged phenotype is established by setting the path coefficients originating from the two phenotypes (and pointing towards delta) to −1 and 1. Formulas for the different heritability estimates are as follows:

h_2 total (unchallenged) $= a_{11}{}^2/(a_{11}{}^2 + e_{11}{}^2)$

h_2 total (challenged) $= (a_{21}{}^2 + a_{22}{}^2)/(a_{21}{}^2 + a_{22}{}^2 + e_{21}{}^2 + e_{22}{}^2)$

h_2 specific (challenged) $= a_{22}{}^2/(a_{21}{}^2 + a_{22}{}^2 + e_{21}{}^2 + e_{22}{}^2)$

h_2 delta (challenged − unchallenged) $= ((a_{21} - a_{11})^2 + a_{22}{}^2)/((a_{21} - a_{11})^2 + a_{22}{}^2 + (e_{21} - e_{11})^2 + e_{22}{}^2)$

of interest. We recently showed that when analyzed as a change score, the heritability of reactivity will reflect an inseparable mix of an amplification or de-amplification of genetic (or environmental) influences already present at rest and newly emerging genetic (or environmental) influences during stress (De Geus et al 2007). Amplified genes are genes that have an effect on individual differences in a cardiovascular trait at rest and these effects become stronger under stress.

Emerging genes are genes that are only expressed during stress and only contribute to the heritability of a cardiovascular trait when it is measured under stressful conditions. In order to distinguish between these different sources of variance we used a conceptually novel approach to the genetics of cardiovascular reactivity, which simultaneously analyses resting and average stress levels in a full bivariate model (De Geus et al 2007).

The averaged cardiovascular response to a choice reaction time and mental arithmetic test was assessed for SBP and DBP, heart rate, pre-ejection period (PEP: an index of sympathetic tone) and respiratory sinus arrhythmia (RSA: an index of heart rate variability [HRV] reflecting parasympathetic tone) in 160 adolescent and 212 middle-aged twin pairs from the Netherlands. Genetic factors significantly contributed to individual differences in resting SBP, DBP, heart rate, PEP and RSA levels in the adolescent (heritability range: 0.31–0.70) and middle-aged (heritability range: 0.32–0.64) cohorts. The effect of these genetic factors was amplified by stress for all variables in the adolescent cohort, and for SBP in the middle-aged cohort. In addition, stress-specific genetic variation emerged for heart rate in both cohorts and for PEP and SBP in the adolescent cohort. Heritability of stress levels of SBP, DBP, heart rate, PEP and RSA ranged from 0.54 to 0.74 in the adolescents and from 0.44 to 0.64 in the middle-aged cohort. On the basis of these results we concluded that exposure to stress uncovers new genetic variance and amplifies the effect of genes that already influence the resting level (De Geus et al 2007). This has clear implications for gene finding studies. The genetic variation that emerges exclusively during stress can only be found in studies that have attempted to measure the levels of the cardiovascular risk factor under challenged conditions as can be induced in a standardized lab setting. Genetic variation that is amplified during stress can be detected using resting levels, but the genetic variance, and hence the power of the study, will be larger if cardiovascular levels are measured during a stressful challenge instead. Capitalizing on the results for RSA we designed a gene-wide candidate gene association study for heart rate variability (HRV) measured at rest and under acute stress as described below.

Identifying genes for heart rate variability

HRV is a simple noninvasive measure of cardiac autonomic function. Reduced HRV, reflecting a shift in cardiac sympathovagal balance from parasympathetic to sympathetic control of the heart rhythm, is a predictor of all-cause mortality, arrhythmic events and sudden death after acute myocardial infarction as well as in the general population (Dekker et al 2000, Tsuji et al 1996). Furthermore, autonomic dysregulation as reflected by lower HRV may play a causal role in the aetiology of essential hypertension as outlined in our gene-environment interaction model of stress-induced hypertension (Imumorin et al 2005).

Previous twin and family studies have found that up to 65% of the variance in HRV at rest can be explained by genetic influences (Boomsma et al 1990, Busjahn et al 1998, De Geus et al 2003, Kupper et al 2004, Snieder et al 1997). However, several lines of evidence suggest that genetic influence may be more pronounced when the subject is challenged by mentally and emotionally taxing tasks (Boomsma et al 1990, Snieder et al 1997). In one study, HRV indexed by RSA was measured in 160 adolescent twin pairs during a rest period and two stressful laboratory tasks. During rest, only 25% of the variance in HRV was accounted for by genetic influences. But under task conditions, the genetic contribution increased to 51% (Boomsma et al 1990). Snieder et al (1997) performed a similar study in 212 middle-aged twin pairs and also observed that the genetic variance of RSA increased from 31% at rest to 43% under stress. Using a multivariate model, we revisited the data sets described above (Boomsma et al 1990, Snieder et al 1997). The results showed that HRV (i.e. RSA) at rest and under stress are influenced by the same genes, but there was a substantial increase in heritability of RSA under mental stress for adolescent twins (from 0.31 to 0.54, $P = 0.001$) and a modest non-significant increase in middle-age twins (from 0.33 to 0.43). That is, the increased heritability under stress was entirely explained by amplification of the influence of genes underlying HRV at rest (De Geus et al 2007). HRV measured during a stressful challenge may, therefore, be the optimal endophenotype for gene finding studies.

Thus, we designed a—currently ongoing—gene-wide candidate gene association study with the objective to identify genes contributing to HRV at rest and during stress. We chose to focus on eight key genes involved in biosynthesis, transport, breakdown and receptor binding of acetylcholine, which is the neurotransmitter of the parasympathetic pathway. Effect of these genes will be assessed by a gene-wide approach with all variants (i.e. tagging SNPs) within a candidate gene considered jointly. We will use data from the TRAILS (Tracking Adolescents' Individual Lives Survey) cohort (Huisman et al 2008), which includes 800 unrelated white youth (mean age: 15.5 ± 0.6) and data on 846 white and 720 black youth (mean age: 18.3 ± 3.3) drawn from the Georgia Cardiovascular Twin Study (Ge et al 2006) and the longitudinal BP Stress cohort. (Dekkers et al 2002). Beat-to-beat heart rate at rest and during several stress tasks has been recorded in all cohorts and can be used to derive HRV. Recent findings from our Georgia Cardiovascular Twin Study have largely replicated our results in the Dutch twins. That is, we found increases in heritability for both our time domain index (RMSSD, root mean square of successive differences) and frequency domain index (HF, high frequency power) of HRV under acute behavioral stress. Heritability of RMSSD increased from 0.48 to 0.58 ($P = 0.07$) and HF increased from 0.50 to 0.58 ($P = 0.20$) from rest to stress. For both parameters, the heritability during stress could largely be attributed to genes that also influence resting levels. Based on all these observations, we expect SNPs with significant effects on HRV to have a larger effect on the challenged phenotype.

A genome-wide analysis scan of systolic and diastolic blood pressure. The effect of gene–stress interaction

Hypertension refers to a clinically significant increase in blood pressure and constitutes an important risk factor for cardiovascular disease. However, identification of risk alleles for hypertension or blood pressure has been notoriously difficult even though the influence of genetic factors has been well-described with heritabilities for SBP and DBP ranging from 40–60% (Snieder 2004). Building on our gene–environment interaction model of stress-induced hypertension (Imumorin et al 2005, Snieder et al 2002), we intend to test the hypothesis that some susceptibility genes for (high) blood pressure only come to expression in (chronically) stressful environments within the context of a genome-wide scan of SBP and DBP.

As part of the Genetic Association Information Network (GAIN) initiative (Manolio et al 2007), 1860 subjects—mainly depressed cases—from the Netherlands Study of Depression and Anxiety (NESDA) (Penninx et al 2008) were genotyped using the 600 k Perlegen chip. All SNPs that successfully pass the genome-wide scan quality control checks will be included in analyses for main genetic effects and gene–stress interactions. Demographical characteristics such as gender, age and socioeconomic status, as well as the severity of depressive symptoms will be used as covariates. In order to limit the already high number of tests involved when analysing gene–stress interaction on a genome-wide scale, we wanted to create a single cumulative index of chronic stress. To this end, we used all available questionnaire- and interview-based measures of exposure to both distal and proximal stress in the entire NESDA sample ($n = 2981$) (Penninx et al 2008).

Available measures of exposure to stress

One distal stressor concerned adverse childhood experiences. We inquired after childhood experiences of psychological neglect, emotional abuse, physical abuse and sexual abuse before age 16 as part of an interview (de Graaf et al 2004). Answers were recorded on a 6 point scale ranging from 'never' to 'very often'. The experiences were summed to provide a total index of childhood adversity.

A more proximal stressor was the experience of serious life events in the 12 months preceding our assessment of blood pressure, as measured by the Brugha Life Events Scale (Brugha et al 1985). Life events were coded as either present or absent and the total number of life events in the prior year indexed the extent to which serious life events were recently experienced.

We also measured perceived lack of social support in the year prior to blood pressure measurement. This was done by an adaptation of the Close Person Questionnaire (Surtees et al 2004), which provides indices of the inadequacy of confiding/emotional support, practical support, and negative aspects of relationships with respondent-nominated close confidants. Respondents could nominate multiple confidants. Following Stansfeld et al (1998) we used ratings of social support of the first close person. The 10 items of this questionnaire were summed for an overall index of the lack of social support. If a respondent was unable to nominate a single close confidant, we imputed the score that corresponded to the 90th percentile for each subscale, respectively, and summed these. This was based on the assumption that the absence of a single close person in one's social network is indicative of severe inadequate social support.

We further assessed whether neighborhood conditions were currently perceived as stressful by means of the sum of four questions that inquired after perceived non-safety, noise created by neighbors or traffic, vandalism or destruction of own property, and overall satisfaction with the neighborhood. Items were scored on a 5-point scale ranging either from 'never' to 'often' or from 'very good' to 'bad'.

To determine the extent to which work was currently perceived as stressful we used a Dutch adaptation of the Job Content Questionnaire (Karasek et al 1998). The five subscales of this questionnaire are: job demands, decision authority, skill discretion, social support at work and job insecurity. Twenty-seven items were rated using a dichotomous format of 'yes' or 'no'. Two items used a 3-point format measuring 'no', 'threat of', or 'actual job loss' in the past year, and the likelihood of job loss ('no', 'somewhat' and 'very likely') in the coming year, respectively. High job demands, low decision authority, low skill discretion, low social support at work and high job insecurity are hypothesized to provide the highest job strain (Karasek et al 1998) and items were summed accordingly into a single index of perceived job stress.

Finally, we used the Daily Hassles Questionnaire (Kanner et al 1981) for a measure of relatively minor life events in the month preceding blood pressure assessment. Respondents were asked to rate the extent to which they had been troubled by 20 different events or situations. Response options were 'not at all', 'somewhat', 'moderately', and 'very much'. The responses to the items were summed into a single index of daily hassles.

Towards a single cumulative index of chronic stress

Table 1 lists the correlation matrix of these six stressors using information from all 2981 NESDA respondents. Prior to analysis, we had expected substantial

TABLE 1   Correlations among available measures of exposure to stress in the NESDA sample*

	Childhood adversity	Life events	Lack of social support	Neighborhood stress	Job stress	Daily hassles
Childhood adversity	1					
Life events	.08	1				
Lack of social support	.15	.05	1			
Neighborhood stress	.16	.06	.10	1		
Job stress	.20	.10	.16	.18	1	
Daily hassles	.29	.10	.29	.23	.36	1

All correlations significantly different from 0 at $P < 0.01$.
* Available sample size is $n = 1679$ for job stress and ranges between $n = 2942$ and $n = 2981$ for the other 5 scales.

correlations given that (i) the majority of our sample ($n = 2608$) have a lifetime diagnosis of a mood or anxiety disorder (i.e. only 373 are healthy controls) and (ii) all measures are retrospective and based on self-report. We expected respondents at risk of anxiety and/or depression to perceive relatively high stress in multiple areas simultaneously. Healthy controls were expected to report low stress overall. Such a pattern would lead to substantial correlations between the different stressors. However, as shown in Table 1, this expected pattern was not confirmed by the data as correlations were only modest in size. Based on our prior expectation of substantial correlations between stressors we had intended to extract their common variance by using factor scores derived from a single factor model. The actual findings suggest that an index based on a simple sum-score of relatively independent stressful experiences is more suitable.

Thus, to calculate the stress index the stressors were summed and divided by six. Note that we calculated standardized scores, which implies that each stressor was equally weighted in the mean sum-score of six stressors. We allowed for one missing value, in which case the sum-score of the stressors was divided by five. In the far majority of cases the missing value concerned work stress, due to the large number of respondents in the sample without a job ($n = 1302$). This index of chronic stress will be used to look for genes underlying SBP and DBP that may only come to expression in stressful environments. It is the hope that such findings may improve insight into the aetiology of hypertension.

Summary and conclusion

Recent progress in variation discovery efforts such as the HapMap have provided a major impetus to gene finding efforts for common complex diseases. These developments have led to two important developments in association studies: the gene-wide candidate gene study and the genome-wide association study. In this chapter, we first described the design of a gene-wide candidate gene association study for HRV that was founded on the observation that exposure to an acute stressful challenge (i.e. gene–stress interaction) produced a more heritable endophenotype. Such a larger heritability potentially facilitates identification of underlying genes. In our second example we showed how the creation of a single cumulative index of chronic stress based on multiple measures of exposure may be applied in a genome-wide association study of blood pressure to find genes that only come to expression in stressful environments. In conclusion, investigation of gene–environment interactions in the context of both gene- and genome-wide association studies may offer important advantages in gene finding efforts for complex traits and diseases.

References

Boomsma DI, van Baal GC, Orlebeke JF 1990 Genetic influences on respiratory sinus arrhythmia across different task conditions. Acta Genet Med Gemellol (Roma) 39:181–191

Brugha T, Bebbington P, Tennant C, Hurry J 1985 The List of Threatening Experiences: a subset of 12 life event categories with considerable long-term contextual threat. Psychol Med 15:189–194

Busjahn A, Voss A, Knoblauch H et al 1998 Angiotensin-converting enzyme and angiotensinogen gene polymorphisms and heart rate variability in twins. Am J Cardiol 81:755–760

Crawford DC, Akey DT, Nickerson DA 2005 The patterns of natural variation in human genes. Annu Rev Genomics Hum Genet 6:287–312

De Geus EJ, Boomsma DI, Snieder H 2003 Genetic correlation of exercise with heart rate and respiratory sinus arrhythmia. Med Sci Sports Exerc 35:1287–1295

De Geus EJ, Kupper N, Boomsma DI, Snieder H 2007 Bivariate genetic modeling of cardiovascular stress reactivity: does stress uncover genetic variance? Psychosom Med 69:356–364

De Graaf R, Bijl RV, ten Have M, Beekman AT, Vollebergh WA 2004 Rapid onset of comorbidity of common mental disorders: findings from the Netherlands Mental Health Survey and Incidence Study (NEMESIS). Acta Psychiatr Scand 109:55–63

Dekker JM, Crow RS, Folsom AR et al 2000 Low heart rate variability in a 2-minute rhythm strip predicts risk of coronary heart disease and mortality from several causes: the ARIC Study. Atherosclerosis Risk In Communities. Circulation 102:1239–1244

Dekkers JC, Snieder H, Van den Oord EJCG, Treiber FA 2002 Moderators of blood pressure development from childhood to adulthood: a 10-year longitudinal study in African- and European American youth. J Pediatr 141:770–779

Frayling TM 2007 Genome-wide association studies provide new insights into type 2 diabetes aetiology. Nat Rev Genet 8:657–662

Ge D, Dong Y, Wang X, Treiber FA, Snieder H 2006 The Georgia Cardiovascular Twin Study: influence of genetic predisposition and chronic stress on risk for cardiovascular disease and type 2 diabetes. Twin Res Hum Genet 9:965–970

Gottesman II, Gould TD 2003 The endophenotype concept in psychiatry: etymology and strategic intentions. Am J Psychiatry 160:636–645

Gould TD, Gottesman II 2006 Psychiatric endophenotypes and the development of valid animal models. Genes Brain Behav 5:113–119

Gu D, Rice T, Wang S et al 2007 Heritability of blood pressure responses to dietary sodium and potassium intake in a Chinese population. Hypertension 50:116–122

Hirschhorn JN, Daly MJ 2005 Genome-wide association studies for common diseases and complex traits. Nat Rev Genet 6:95–108

Huisman M, Oldehinkel AJ, De Winter AD et al 2008 Cohort profile: The Dutch 'TRacking Adolescents' Individual Lives' Survey'; TRAILS. Int J Epidemiol. In press

Imumorin IK, Dong Y, Zhu H et al 2005 A gene–environment interaction model of stress-induced hypertension. Cardiovasc Toxicol 5:109–132

International HapMap Consortium 2005 A haplotype map of the human genome. Nature 437:1299–1320

Kanner AD, Coyne JC, Schaefer C, Lazarus RS 1981 Comparison of two modes of stress measurement: daily hassles and uplifts versus major life events. J Behav Med 4:1–39

Karasek R, Brisson C, Kawakami N, Houtman I, Bongers P, Amick B 1998 The Job Content Questionnaire (JCQ): an instrument for internationally comparative assessments of psycho-social job characteristics. J Occup Health Psychol 3:322–355

Klein RJ, Zeiss C, Chew EY et al 2005 Complement factor H polymorphism in age-related macular degeneration. Science 308:385–389

Kupper NH, Willemsen G, van den Berg M et al 2004 Heritability of ambulatory heart rate variability. Circulation 110:2792–2796

Manolio TA, Rodriguez LL, Brooks L et al 2007 New models of collaboration in genome-wide association studies: the Genetic Association Information Network. Nat Genet 39:1045–1051

Neale BM, Sham PC 2004 The future of association studies: gene-based analysis and replication. Am J Hum Genet 75:353–362

Obrist PA, Black AH, Brener J, DiCara LV 1974 Cardiovascular psychophysiology. Chicago: Aldine Publishing Company

Penninx BWJH, Beekman ATF, Smit JH et al 2008 The Netherlands Study of Depression and Anxiety (NESDA): rationale, objectives and methods. Int J Methods Psychiatr Res, in press

Schork NJ 1997 Genetically complex cardiovascular traits: origins, problems, and potential solutions. Hypertension 29:145–149

Sing CF, Stengard JH, Kardia SL 2003 Genes, environment, and cardiovascular disease. Arterioscler Thromb Vasc Biol 23:1190–1196

Snieder H 2004 Familial aggregation of blood pressure. In: Portman RJ, Sorof JM, Ingelfinger JR (eds) Clinical hypertension and vascular disease: pediatric hypertension. Totowa, NJ: Humana Press, p 265–278

Snieder H, Boomsma DI, Van Doornen LJ, De Geus EJ 1997 Heritability of respiratory sinus arrhythmia: dependency on task and respiration rate. Psychophysiology 34:317–328

Snieder H, Harshfield GA, Barbeau P, Pollock DM, Pollock JS, Treiber FA 2002 Dissecting the genetic architecture of the cardiovascular and renal stress response. Biol Psychol 61:73–95

Stansfeld SA, Fuhrer R, Shipley MJ 1998 Types of social support as predictors of psychiatric morbidity in a cohort of British Civil Servants (Whitehall II Study). Psychol Med 28:881–892

Surtees PG, Wainwright NW, Khaw KT 2004 Obesity, confidant support and functional health: cross-sectional evidence from the EPIC-Norfolk cohort. Int J Obes Relat Metab Disord 28:748–758

Tsuji H, Larson MG, Venditti FJ Jr et al 1996 Impact of reduced heart rate variability on risk for cardiac events. The Framingham Heart Study. Circulation 94:2850–2855

Wang WY, Barratt BJ, Clayton DG, Todd JA 2005 Genome-wide association studies: theoretical and practical concerns. Nat Rev Genet 6:109–118

DISCUSSION

Stankovich: I like the paper by Neale & Sham (2004) that you referred to, in which they advocate taking a whole gene approach and correcting for the number of tests that you have to conduct within one gene. You suggested that this might be a reason for the failure to replicate. I am not so convinced by that argument: if the particular polymorphism is strongly associated with disease in one population, why do you need a replication study to look at the whole gene? Could potential environmental heterogeneity be a reason for arguing that there would be different polymorphisms in the gene associated with different populations, or is it adequate just to conduct the replication study on the previously associated polymorphism?

Snieder: One issue is, are you sure that the SNP where you find the signal is the causal SNP? If you have a lot of evidence that it is a causal SNP, you might as well focus on that causal SNP. If you don't know there may be an advantage in looking at the whole gene first to see whether you are able to replicate that this particular gene is involved in the phenotype you are interested in. Then afterwards you can try to narrow it down to certain polymorphisms.

Martin: Your message is an interesting one. We have two different models of sex limitation. We talk about the scalar sex limitation, and you are really talking about scalar gene × environment (G × E) interaction where you are just seeing the same genes amplified in effect under a stress. Analogous with sex limitation, there is also non-scalar sex limitation where quite different genes are turned on. It was interesting in your case that you are not seeing this. I suppose the first example I came across of this was when we did our alcohol challenge study. We trained a bunch of twins on psychomotor tasks when they were sober and then got them drunk and measured their psychomotor reactivity. After this we found that their heritability went up, because new genes were being switched on. This is an extreme example, but I wanted to mention it as a counter that one does need to be a little careful with this model to make sure you are switching on the same genes.

Heath: What was your 95% confidence interval (CI) on the genetic correlation between the conditions? You gave us the estimates, conditional on the assumption that you know with precision that there is perfect genetic correlation. But if you relax that assumption, what is your confidence interval on your genetic correlation?

Snieder: You mean if we don't look within our best fitting model but in a more general model? I would have to look it up. The fact that it was replicated in a completely different sample and in a different country is encouraging.

Heath: Your 95% CI could still be from 0.01 to 1.00. It is important to know this.

Martin: Equally, your 95% CI on the specific, which you drop from the model; it would be interesting to know how large that might be, and what sample size you would need to detect it.

Snieder: We did look at blood pressure and heart rate, and found some new genes emerging for heart rate, for example, in both middle-aged and adolescent cohorts.

Heath: P values alone are not enough.

Snieder: There might still be some small specific genetic effects that we have missed.

Battaglia: I have a comment about the emergence in your heart rate variability. There might be a way in which you can widen the difference between the baseline and elicitation. This is by measuring heart rate variability during non REM sleep, on the basis of spontaneous movements. When we sleep in stages 3 and 4 we move spontaneously, which causes labor, which increases the heart rate. This is a neat measure separate from any emotional influence. My expectation is that with this, the variability among individuals will widen from baseline to stimulus, and might lead you to find more genes.

Snieder: We are actually not interested in new genes emerging! We were happy to find that the same genes have an effect on the stress, as those that have an effect at rest. It is just that the heritability is much larger, so it should be easier to find these genes if you use a different endophenotype. This is the point I was trying to make.

Kleeberger: I have a couple of questions about the way that heart rate variability was measured. Was ventilation considered? I believe this can have an effect on heart rate variability.

Snieder: In the original cohort, we measured respiration rate. We did a multivariate model where we included respiration rate. We found it doesn't change the heritability much.

Kleeberger: You used high frequency in your heart rate variability measurements. Is low frequency not a good predictor? Is total power a good predictor?

Snieder: We have all those measures as well. It is just that we decided to look at high frequency because that seems to reflect vagal tone, which is the best measure of parasympathetic nervous system activity. The other frequency components are more of a mixture of sympathetic and parasympathetic activity.

Kleeberger: Was the stress you used behavioral and social stress?

Snieder: Yes, they are mental arithmetic or video tasks, or a social interview type of setting where they have to recall an event that made them angry. We were averaging the response over all those stresses. We weren't interested in specific stresses.

Kleeberger: If you looked at physiological stress, or additional kinds of stresses, would you have the same sorts of relationships?

Snieder: One test we did was the cold pressor test, which you could view as more of a physical stressor. It didn't elicit any response in terms of heart rate variability.

Uher: Doesn't it decrease heart rate?

Snieder: It has a huge effect on blood pressure, which increases.

Kleeberger: We have performed a genome-wide association study in mice to identify QTLs and candidate genes for heart rate variability, and we have identified some interesting candidate genes (Howden et al 2008). These are mice that are instrumented for radiotelemetry measurements of heart rate. Like you, we find considerable variability (across strains in this case), but it doesn't necessarily predict how they respond to an oxidant stress, for example.

Heath: I am contemplating the varieties of stress, and whether you think the questionnaire-based measures that you are using will show the same relationships as the experimental measures of stress. In one you are looking at what are potentially effects over a long time, and in the other you are looking at short-term effects.

Snieder: Yes, they are totally different types of measurements. With the genome-wide association study you have to work with the measurements you have. In this case, exposure to stressful events was measured rigorously, in many different ways. You can still argue that it is not optimal. There is a lot of measurement error and it is all self-report. This is why we try to combine all these different measures into a single index.

Heath: It would be nice to have a model system in which you have a candidate gene which you know has effects, and you can look for moderating effects of your hypothesized environmental measure before you get into the business of correcting for 500 000 SNPs.

Rutter: A major problem for environmental researchers is the development of measures that can be used on large populations. Not only do you need measures that are sensitive to what it is you are interested in, but also you need to precede this by showing that the effects are both significant and environmentally mediated. I would be pretty skeptical about these kinds of questionnaire measures as being good enough for the job you are trying to use them for. I have a different question. You started your presentation by arguing that the challenge findings in themselves were a reflection of a gene–environment interaction. That is to say, the environmental challenge is bringing out a genetic effect. I was puzzled by the end of your presentation, where you thought the important thing is raising heritability, which I would have thought is the least interesting bit. If you raise heritability there are many reasons why it might go up, but it is not obvious that this is a useful goal in itself. So why the shift?

Snieder: If you are interested in gene finding, then heritability is important. The first thing you do is look at the twin and family literature to see whether there is any heritability at all. Heritability in itself isn't that important. With the type of models that I used you can break it down into the reasons why the heritability might go up or down, and what it consists of: the amplification of the same genes or the emergence of novel genetic effects. All these types of effects constitute a type of G × E interaction. It is the environment that is changed experimentally.

Martin: What you hope is that taking a standard Fisher or Falconer model, (where m = mean, d = additive deviation, and h = dominance deviation) whether with your single genotype you actually expanding the size of d, the additive deviation, to increase your power. I guess this is a reasonable hope if the heritability is going up and the environmental variance going down.

Rutter: In effect, the level of heritability has proved to be a pretty hopeless guide to finding genes.

Martin: In general this is true, but in this case, if he is right that there are no new genes being switched on and he is amplifying the effect size of the genes that are there, it is a reasonable premise.

Rutter: It's an interesting finding, but I'd just interpret it another way. It is using something that is reflecting G × E interaction rather than something that is increasing heritability.

Martin: This was my initial point. It is G × E interaction in the sense of being scalar as opposed to being non-scalar. Most of us are assuming we are talking about the non-scalar case of new genes being switched on, whereas here we are talking about the scalar case.

References

Howden R, Liu E, Miller-DeGraff L et al 2008 The genetic contribution to heart rate and heart rate variability in mice. Manuscript in revision

Neale BM, Sham PC 2004 The future of association studies: gene-based analysis and replication. Am J Hum Genet 75:353–362

6. Practice and public policy in the era of gene–environment interactions

Kenneth A. Dodge

Center for Child and Family Policy, Duke University, Durham, North Carolina, USA

Abstract. This chapter argues that implications of the gene–environment interaction revolution for public policy and practice are contingent on how the findings get framed in public discourse. Frame analysis is used to identify the implications of the ways in which findings are cast. The frame of 'defective group' perpetuates racial and class stereotypes and limits policy efforts to redress health disparities. Furthermore, empirical evidence finds it inaccurate. The frame of 'defective gene' precludes the adaptive genetic significance of genes. The frame of 'individual genetic profile' offers individualized health care but risks misapplication in policies that place responsibility for disease prevention on the individual to the policy relief of industry and toxic environments. Framing the interaction in terms of 'defective environments' promotes the identification of harmful environments that can be regulated through policy. The 'therapeutic environment' frame offers hope of discovering interventions that have greater precision and effectiveness but risks dis-incentivizing the pharmaceutical industry from discovering drug treatments for 'obscure' gene–environment match groups. Can a more accurate and helpful framing of the gene–environment interaction be identified? Findings that genes shape environments and that environments alter the gene pool suggest a more textured and symbiotic relationship that is still in search of an apt public framing.

2008 Genetic effects on environmental vulnerability to disease. Wiley, Chichester (Novartis Foundation Symposium) p 87–102

We are in the midst of a revolution in science and are about to enter a revolution in health care and social policy. Sweeping statements, such as by Khoury et al (2000), are energizing scientists and entering the vocabulary of the public and policy makers: '*all* human disease is the result of interactions between genetic variation and the environment (broadly defined to include dietary, infectious, chemical, physical, and social factors).'

Translation to practice and policy

What are implications of this revolution for practice and public policy? We can be tempted to leap to the exciting, or even Orwellian, possibility of genetic screening that could tell someone at birth when he or she is likely to die and of what cause, contingent on various environmental conditions, and how to intervene to live longer.

In the USA today, the discovery of the breast cancer genes (*BRCA1* and *BRCA2*) that raise the probability of developing breast cancer to 75% have led to unprecedented screening and preventive surgery. Among women screened positive for *BRCA1* or *BRCA2*, fully one-third have opted for preventive mastectomy and over half for ovary removal (Harmon 2007). The numbers of women seeking screening has doubled in the past two years and is expected to double yet again this year. Should private health insurance or government healthcare pay for screening? Should it pay for preventive surgery in an otherwise healthy woman? As we contemplate these questions, public debates range from proclamations that mapping the human genome will lead to the prevention of all disease to Luddite worries about genetic hegemony (Shostak 2000).

The thesis of this chapter is that the policy impact of this revolution will depend on how the empirical findings and concepts are framed in public discourse. Thus, the scientific field will need to move slowly, carefully, and cautiously to make sure to fulfill its responsibility to 'get it right' in framing because the consequences will be huge.

The scholarly study of public policy reveals the importance of how issues get framed (intentionally or implicitly) through metaphors, stories, and narratives in determining policy outcomes. Frame analysis is based in Bateson's (1972) studies of culture in anthropology, followed by Lakoff's (1996) analyses in linguistics. In cognitive psychology, Schank (1998) and Tversky & Kahneman (1986) have demonstrated that humans use heuristics to understand and act on phenomena. More recently, Goss (2006) has brought framing theory from political science to bear on understanding how policy issues such as gun control and women's rights are deliberated in policy debates.

The study of frames (Dodge 2008) reveals that they are most effective when they first enable the listener to *assimilate* the concept into existing modes of understanding, and then *accommodate* the concept to a new and different understanding than previously. So, framing a new finding as indicating that *genetic causes have been revealed through the effects of drugs* casts the phenomenon as supporting perhaps inevitable genetic causes of a disease. In contrast, framing the finding as indicating that particular environments can prevent disease from ever occurring casts the phenomenon in a very different light. Next, an effective frame leads the listener to take *action,* such as policy or intervention. One frame might lead to support for genetic screening, whereas another framing of the same phenomenon might lead to environmental engineering. But the test of the utility of a frame will also depend on its *accuracy,* and it is here that scientists have a responsibility. Scientists cannot simply report their findings;

they are also responsible for how those findings are heard. The remainder of this chapter addresses different framings of the gene-by-environment interaction.

The 'defective' race or group

The notion of a defective group of human beings continues to haunt both science and public discussion in fields as diverse as intelligence and violent behavior. Nobel Laureate James Watson lost his prestigious position following his statement that he was 'inherently gloomy about the prospect of Africa' because 'most tests' revealed that persons of African descent are not as intelligent as other groups (Sunday Times 2007). Watson had implicitly asserted a gene-by-environment interaction framed in a way that implies that one genetic group (Africans) is not only defective at birth but also defective in being less responsive to environmental stimulation than other genetic groups, leading him to be gloomy about prospects for public policies that would seek to enhance outcomes for persons of African descent.

A similar sentiment was expressed by Herrnstein and Murray in *The bell curve*: 'Much of public policy toward the disadvantaged starts from the premise that interventions can make up for genetic . . . disadvantages, and *that premise is overly optimistic* . . . Much can and should be done . . . for those who have the *greatest* potential' (Herrnstein & Murray 1994, p 550). Again, the gene-by-environment interaction is invoked and framed as indicating that a disadvantaged group is not only defective in capacity but also unable to benefit from environmental stimulation.

Several consequences ensue from this framing. First, at stake is public spending directed toward various groups. Duster (1990) has argued that a framing on genetic risk will reduce policy pressure to intervene environmentally to enhance outcomes for disadvantaged African-American school children growing up in dangerous neighborhoods. Di Chiro (2002) argued that this same policy direction will also reduce efforts to identify environmental causes of adverse outcomes for African-Americans. Second, public trust in science may be compromised. The fear in the USA that scientists will disrespect specific minority groups, and African-Americans in particular, has become so great, that research on gene–environment interaction has been hampered by low rates of participation by African-Americans. Dodge (unpublished work 2007) has shown that, although consent rates for participation in research do not vary across ethnic groups, participation in cheek-swab DNA collection is significantly lower among African-Americans. Furthermore, ethics boards that review human subject research have become extremely cautious in requiring cumbersome consent procedures and protective measures that inhibit participation.

In order to regain the public trust, scientists must be accurate in framing the shape of the interaction effect. Most scholars of gene-environment interaction effects in violent behavior have framed the effect in terms of the magnitude of the genetic effect under different environmental conditions (Dodge & Sherrill 2006). What would happen if the interaction were framed in terms of the magnitude of the environmental effect under different genetic conditions? Mednick & Christiansen (1977) proposed that the group at *highest* genetic risk for violence would display the *smallest* environmental effect, that is, they would be impervious to environmental impact and would be destined for violent outcomes under any conditions. This pessimistic hypothesis casts the fate of a high-risk group in much the same terms as that proposed by Watson and Herrnstein and Murray.

Dodge & Sherrill (2006) suggested the opposite hypothesis that it would be the group at highest genetic risk that would show the *strongest* environmental response to an adverse environment. It turns out that this shape characterizes most empirical findings in violence. As depicted in Fig. 1, findings from the Environmental Risk study of twins in the UK by Jaffee et al (2005) indicate that children are highest heritable risk demonstrate the steepest environmental gradient. The psychiatric diagnosis of one's twin can be used as a proxy for one's heritable risk for conduct disorder, and that factor can be crossed with one's early life history of physical maltreatment, which is the most potent environmental toxin for conduct disorder.

Similarly shaped interaction effects have been reported by Cloninger et al (1982) in a study that was among the first to report the heritability by environment interaction effect. The findings decidedly showed that the environment had the strongest impact on the group at highest heritable risk, although Cloninger emphasized

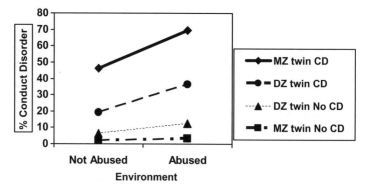

FIG. 1.   The interaction between heritable risk and environmental physical maltreatment in predicting psychiatric conduct disorder. Data from Jaffee et al (2005).

the heritable risk framing. Caspi et al (2002) have also shown that children at genetic risk for violence due to a polymorphism in the monoamine oxidase A (*MAOA*) gene are *more* affected by physical maltreatment than are children without this polymorphism. Dodge & Sherrill (2006) have identified 30 published studies reporting some form of person-by-environment interaction effect in conduct disorder, and all 30 have yielded a pattern in which the environmental effect is larger among the group at high genetic or person risk than among the group at low genetic risk.

Several implications follow from these findings. First, it can be hypothesized that a primary mechanism for aggressive behavior is a deficient response to a threatening social environment. Faced with severe and chronic threat, some individuals develop a vulnerability to emotional and behavioral dysregulation and either hyper-react with violence or hypo-react with emotional blunting and psychopathy. Perhaps this environmental effect tells us something about the nature of the genetic risk itself, that is, a propensity to be highly sensitive to environmental stressors. This hypothesis might be framed as having 'thin skin'; that is, the tendency to be emotionally unsteady and 'prickly' in response to threat.

The second implication runs counter to common assumptions about clinical practice that 'psychopathic individuals are extremely difficult to treat, if not immune to treatment' (Salekin 2002, p 79), because the findings suggest that these individuals are highly susceptible to environmental impact, at least early in life.

Finally, for public policy, the findings suggest that *more* resources should be directed toward high-biological risk groups, including infants with low birth weight or high birth complications, children of prisoners, and maltreated children.

The defective gene

Nebert (1999 p 245), framed the emergent field of ecogenetics as 'the study of genetically determined variations that are revealed solely by the effects of drugs'. Note the unimportance of the environment except to *reveal* the genetic cause (not co-shape it, but to reveal it). The implication is that, once 'defective' genes are identified, environmental ways could be discovered to knock out these genes or select against them across generations. This possibility is exciting, but it ignores the possibility of adaptive genetic significance. Genes that may be related to adverse outcomes under certain environmental conditions may also be related to particularly favorable outcomes under other environmental conditions. Garmezy (1991) heralded this hypothesis, but few empirical tests of this notion have provided support for this hypothesis. One example of this cross-over interaction comes from the author's longitudinal study, the Child Development Project.

Goodnight et al (2006) have identified a trait called reward dominance that predicts adverse outcomes such as substance abuse, violence, and risky sexual behavior. The trait is conceptualized as an overactive behavioral activation system (Gray 1982), in which the individual is highly attentive and responsive to rewarding stimuli to the neglect of attention to punishment and inhibitory factors. It is measured through experimental laboratory gambling tasks in which participants betting performance is mapped following histories of experimenter-manipulated wins and losses. Goodnight et al (2006) have found that adolescents who are reward dominant are likely to become increasingly antisocial, but only when they are also exposed to deviant peers who reward antisocial behavior. Should we conclude that reward dominance is a characteristic that should be selected out? Is this a 'bad gene'?

In a separate study, Goodnight et al (unpublished paper 2007) found that reward dominant adolescents who are exposed to high rates of rewarding and positive parenting are likely to become *less* antisocial than non-reward dominant adolescents exposed to similar rewarding parenting. That is, the reward dominant adolescents seem to be especially equipped and capable of taking advantage of positive parenting. Together, these two studies demonstrate that reward dominant adolescents can become either increasingly antisocial or less antisocial, depending on the reward structure in their environment. Reward dominance is not inherently good or bad, but its impact depends on the environmental context in which the person lives. The possibility of this form of adaptive genetic significance serves as a caution to rash attempts to cast genetic risk in one environmental context as indicating a 'defective gene'.

The individual genetic profile

Gene–environment interactions have the potential to identify an individual's likelihood of developing a disorder in response to environmental circumstances and the probability of responding favorably or unfavorably to therapies for disease. The future of healthcare policy might include broad genetic screening, individual profiling, and individualized prevention and treatment. Is there any downside to this future?

One domain in which genetic screening can be used already in health care concerns smoking cessation. The transdermal nicotine patch has been identified as an effective devise for cigarette smoking cessation, but its effectiveness is limited to individuals with particular clusters of single nucleotide polymorphisms (SNPs) whose allele frequencies distinguish successful abstainers from failed abstainers. This work suggests that the day is near when genetic screening could be used to match individuals with particular therapies.

However, the tobacco industry might alternately frame this work as indicating individual responsibility for the development or treatment of a disorder. Peterson & Lupton (1996) noted that individualizing strategies for disease prevention could empower individuals to assume responsibility for their own health outcomes but could also render them legally and finally responsible for their diseases. Proctor (1995, p 107) commented that when genotyping becomes readily accessible 'one could plausibly argue that people who come down with the disease have at least partly their own heredity to blame.' The implications for litigation are enormous. In the USA, tobacco companies have been held liable for lung cancer deaths and have been forced to pay billions of dollars in settlements. Is it plausible that a future jury could make the inference that if a gene is partly responsible for a disease outcome, then the person is responsible and the environment, in this case, the tobacco industry, is less responsible?

Shostak (2000) has suggested that assigning responsibility to the individual could also be used as a way to avoid environmental change. Instead of requiring industries to clean up toxic environments, governments might require individuals to self-select environments. Instead of requiring a factory owner to keep toxic substances away from its employees, government might endorse genetic screening to select individuals for employment, allowing the employer to discriminate against individuals at highest risk for adverse reactions to toxic substances. Likewise, a workman's compensation claim might be reduced in a situation for which a genetically vulnerable person claims that an employer is responsible for his or her disease.

Lippman (1992) summarized the conflict as follows: 'Recently, a genetic variation said to be associated with increased susceptibility to lead poisoning was described in the literature, with the authors implying that this might be a useful objective for a screening program. Do we really want to screen for genes rather than clean out lead to prevent the avoidable damage known to affect the millions of children unnecessarily exposed annually to this toxic agent?'

A general concern here is that a focus on individual genetic susceptibility might shift public policy formulation to the individual level and away from social, economic, environmental, and political factors in disease and its prevention, remediation and compensation.

The defective environment

The defective environment frame seems more straightforward. It is difficult to refute the rationality or empirical findings that ecological disadvantage, in the form of severe poverty, child abuse, extreme stress, or chemical exposure, is harmful to development. The fact of gene-by-environment interactions and the finding that certain groups of individuals are able to emerge from adversity unscathed does not cast doubt on the fact that toxic environments cause harm for at least some of the human species.

But is it possible that interaction findings could weaken the case of environmentalists by restricting the risk to subgroups, especially disadvantaged subgroups? Is it also possible that some seeming defective environments will prove to be therapeutic for other persons with regard to other outcomes?

The environment as light switch operator

The environment can also be framed as the 'light switch operator.' Although this frame places the ultimate mechanism in the switch, that is, genes, it frames the role of the environment as the controller of that switch. Working with rat pups, Jirtle & Skinner (2007) have identified environmental exposures, including nutrients and toxins, that trigger methylation, which acts as a 'lock' on the switch, or perhaps 'gum' which covers the switch, preventing it from operating. In this way, the environment controls long-term effects on gene action. Meaney (2001) has shown that maternal mal-nurturing in rats (failure to lick the pup) leads to loss of methylation in the glucocorticoid receptor gene which results in higher levels of stress hormones as an adult. Furthermore, they (Szyf et al 2005) have found that the process is reversible by injection of various amino acids. Barker et al (2005) have argued that poor fetal growth, linked to the environmental factor of prenatal care, 'wires' genes to respond differently to future environments in ways that potentiate disease outcomes such as coronary heart disease and diabetes. It is a form of gene-by-environment-by-environment interaction. This framing of gene-by-environment interplay offers hope of finding environmental interventions that can turn on and off the light switch to gene expression.

The therapeutic environment

Finally, the gene-by-environment interaction suggests that the environment can be framed as therapeutic. A child at risk for conduct disorder because of a polymorphism in *MAOA* may be protected from that disorder if the early environment can be free from maltreatment (Caspi et al 2002). A chronic smoker with a particular genetic profile might be a strong candidate for therapy via a transdermal nicotine patch (Uhl et al 2007). Early maternal care, in the form of a rat mother licking her pup, can prevent that pup from stress disease outcomes (Szyf et al 2005). These are genuinely prophylactic or therapeutic environments, and this is the hope and promise of the current revolution. However, is it plausible that an environment that is therapeutic for one outcome might prove to be harmful for other individuals or for other outcomes? Drinking several glasses of wine a day might prove prophylactic for cardiovascular outcomes for some persons but harmful for other outcomes for other persons.

Although the precision and tailoring of the environmental intervention may be enhanced by gene-environment interplay, one policy concern is that developers of interventions, particularly pharmaceutical companies, may become less incentivized to develop drug treatments for a more precisely matched but smaller proportion of the population. A modestly effective drug that could be sold to 20% of the population with a general disorder such as hypertension will achieve more sales than another drug that is targeted at only 1% of the population, even though the latter drug-population match may be far more effective. Similar problems have been identified in finding treatments for so-called obscure disorders. Gene–drug interactions suggest that treatments may be more obscure. So a major policy question is how to induce drug companies, as well as developers of behavioral treatments, to invest in identifying treatments for smaller and smaller segments of the population.

Toward more accurate and helpful framing

How can gene–environment interplay be re-framed in ways that are less likely to lead to unwarranted public reactions and are more accurate? One problem with all of these frames is that they fail to depict the dynamic quality of the interaction effect. They are framed in ways that hold one factor constant while emphasizing the magnitude of the effect of the other factor, as if genes and environments passively exist independently of each other. This static framing of interactions is how Sir Ronald Fisher (1918) understood the world as a statistician in the early twentieth century. Lancelot Hogben (1932) also articulated the gene-by-environment interaction in 1932, but in a more dynamic way. Following Hogben, the emergent scientific understanding is that the gene-environment interaction is not merely a passive statistical phenomenon but indicates the fact that genes are always contextualized in environments and vice-versa. The relationship is one of figure and ground. Like the picture of two faces and a goblet, one cannot frame a genetic effect without contextualizing that effect within an environmental context, and one cannot frame an environmental effect without grounding that understanding in a population with assumed genetic parameters. They assume each other in a symbiotic relationship. With this more textured framing, a radically different, and perhaps more accurate, understanding of how disease develops can emerge.

Conclusion

There are many reasons to be excited about the revolution that is upon us. The eugenics of screening and matching an environmental treatment to a genetic profile is undeniably exciting. But there are also numerous reasons to move cautiously. Gene–environment interactions might turn out to be disguises for gene–gene interactions. Measurement of the environment is woefully inadequate and needs to be made more precise, just as most drugs are blunt instruments in need of improved precision. Limited range of measurement might obscure main effects through ceiling and floor measurement limits. And, most importantly, the way that the interaction effect is framed has reverberating consequences, many of which may be unintended.

Although the eventual outcome of this scientific revolution may be to change completely the way that health care and social services policy are implemented, for the moment the implications seem to be more obvious for science than for policy (Rutter et al 2006). These findings are revolutionizing the way that we understand the processes and mechanisms through which disease develops.

References

Barker DJP, Osmond C, Forsen TJ, Kajantie E, Eriksson JG 2005 Trajectories of growth among children who have coronary events as adults. New Engl J Med 353:1802–1809

Bateson G 1972 Steps to an ecology of mind: collected essays in anthropology, psychiatry, evolution, and epistemology. University of Chicago Press, Chicago, IL

Caspi A, McClay J, Moffitt TE et al 2002 Role of genotype in the cycle of violence in maltreated children. Science 297:851–854

Cloninger R, Sigvardsson S, Bohman M, von Knorring A 1982 Predisposition to petty criminality in Swedish adoptees. Arch Gen Psychiatry 39:1242–1247

Di Chiro G 2002 A new biotechnological 'fix' for environmental health? Examining the environmental genome project. Women and Environments International Magazine 52/53

Dodge KA 2008 Framing public policy and prevention of chronic violence in American youth. Amer Psych, in press

Dodge KA, Sherrill MR 2006 Deviant peer-group effects in youth mental health interventions. In: Dodge KA, Dishion TJ, Lansford JE (eds) Deviant peer influences in programs for youth: problems and solutions. New York: Guilford, p 97–121

Duster T 1990 Backdoor to eugenics. New York: Routledge

Fisher R 1918 The correlation between relatives on the supposition of Mendelian inheritance. Trans R Soc Edinb 52:399–433

Garmezy N 1991 Resiliency and vulnerability to adverse developmental outcomes associated with poverty. Am Beh Scientist 34:416–430

Goodnight JA, Bates JE, Newman JP, Dodge KA, Pettit GS 2006 The interactive influences of friend deviance and reward dominance on the development of externalizing behavior during middle adolescence. J Abnorm Child Psychol 34:573–583

Goss KA 2006 Disarmed: the missing movement for gun control in America. Princeton University Press, Princeton, NJ

Gray JA 1982 The neuropsychology of anxiety: an enquiry into the function of the septohippocampal system. Oxford University Press, Oxford

Harmon A 2007 Cancer gene carriers weigh options. New York Times, 16 Sept 2007

Herrnstein RJ, Murray C 1994 The bell curve. Free Press

Hogben L 1932 Genetic principles in medicine and social science. New York: Knopf

Jaffee SR, Caspi A, Moffitt TE et al 2005 Nature x nurture: genetic vulnerabilities interact with physical maltreatment to promote conduct problems. Dev Psychopathol 17:67–84

Jirtle RL, Skinner MK 2007 Environmental epigenomics and disease susceptibility. Nat Rev Genet 8:253–562

Khoury MJ, Burke W, Thompson EJ 2000 Genetics and public health in the 21st century: using genetic information to improve health and prevent disease. Oxford University Press, Oxford

Lakoff G 1996 Moral politics: what conservatives know that liberals don't. University of Chicago Press, Chicago, IL

Lippman A 1992 Prenatal genetic testing and screening: constructing needs and reinforcing inequities. Am J Law Math 17:15–50

Meaney MJ 2001 Maternal care, gene expression, and the transmission of individual differences in stress reactivity across generations. Ann Rev Neurosci 24:1161–1192

Mednick SA, Christiansen KO 1977 Biosocial bases of criminal behavior. Gardner Press

Nebert DW 1999 Pharmacogenetics and pharmacogenomics: why is this relevant to the clinical geneticist? Clin Gen 56:247–258

Peterson A, Lupton D 1996 The new public health: health and self in the age of risk. Sage

Proctor RN 1995 Cancer wars: how politics shapes what we know and don't know about cancer. Basic Books

Rutter M, Moffitt TE, Caspi A 2006 Gene–environment interplay and psychopathology: multiple varieties but real effects. J Child Psychol Psychiatry 47:226–261

Schank R 1998 Tell me a story: narrative and intelligence. Northwestern University Press

Salekin RT 2002 Psychopathy and therapeutic pessimism: clinical lore or clinical reality? Clin Psychol Rev 22:79–112

Szyf M, Weaver IC, Champagne FA, Diorio J, Meaney MJ 2005 Maternal programming of steroid receptor expression and phenotype through DNA methylation in the rat. Front Neuroendocrinol 26:139–162

Shostak S 2000 Locating gene-environment interaction: at the intersections of genetics and public health. Soc Sci Med 56:2327–2342

Sunday Times 2007 The elementary DNA of Dr. Watson, Sunday Times 14 Oct 2007

Tversky A, Kahneman D 1986 Rational choice and the framing of decisions. J Business 59: No.4, Pt.2.

Uhl GR, Liu Q-R, Drgon T, Johnson C, Walther D, Rose JE 2007 Molecular genetics of nicotine dependence and abstinence: whole genome association using 520,000 SNPs BMC Genet 8:10

DISCUSSION

Uher: You mentioned the crossed interactions, where the same people would be sensitive to both negative and positive environments so that there is no main effect of the gene; the gene is neither bad nor good. You mentioned that such crossed interactions are probably rare.

Dodge: I meant not that these people are rare, but that these findings are rare. That is, it is rare to find a person variable in which the same factor is associated with good outcomes under some conditions and bad outcomes under others. An excellent discussion of the hypothesis that some children are highly susceptible to environmental impact, both positively and negatively, has been asserted recently by Belsky and his colleagues (Belsky et al 2007).

Uber: Are they likely to be rare in reality as causal mechanisms of disorders, or are they rare because they are more difficult to find, because we select our candidates on the basis of main effects?

Dodge: The latter option is a possibility. There is very little genetic study of positive outcomes.

Uber: It makes sense from an evolutionary perspective that genes will be maintained because they have adaptive significance.

Dodge: Yes. Except in rare cases these genes have been maintained over evolutionary time. They may be common. They may have some functional significance. So we need to be slow to do what the public will want to do, which is to knock out those genes through feticide or some other means.

Martinez: There is a problem we face in order to do what you are asking us to do. This is to try to explain to people who don't do this for a living what it is that we are doing. The first problem is that we have to work using metaphors. We can't explain exactly what we are doing directly. We have to explain it with words that have no biological meaning. Genes don't 'control' anything: how can they control anything if they are inert pieces of matter away from a cellular context. But we all use the terms 'control', 'regulation' or 'program'. This is difficult to demonstrate biologically. The first problem we have is how to say the things that we do.

Dodge: It is simultaneously helpful: we can assimilate it, accommodate it, it leads us to action. It is also dangerous to the extent that it is not accurate.

Martinez: Sometimes we eat up our own metaphors. We start using them in ways that have nothing to do with what the reality is. You called it 'frames'; I prefer to call it 'metaphors'.

Dodge: A metaphor is a specific kind of frame.

Martinez: This is mostly what people talk about: the gene 'for'; it's 'in our genes'. People like this because it is the way they tend to think of things in life, with some ultimate cause. The second issue is linearity. People tend to think linearly about things. There are many more of these crossing interactions than people realize. As soon as we start searching for them (and the problem we have is that we don't search for them; if we don't see the main effect we say there is nothing there) we start seeing things that are non-linear. These are difficult to explain to people, because how can something that is 'good' in one context be 'bad' in another? A recent paper showed (Flegal et al 2007) that being mildly overweight is good for you. It happens that mortality is lower in overweight, but not obese, people. The curve of mortality is U-shaped. So why have we been encouraging people with BMIs of 25–30 to diet to death? Because we tend to think in a linear way. We also have a natural tendency to think that only strong effects and strong links matter. In other words, we think it is only big effects that matter. People have problems with weak-linking effects in causality, but these are probably the only ones that make evolution possible. Strong links are indestructible and you can't evolve. The issue of selection for evolvability—selection for weak links—is an important one. Perhaps the majority of links we will be studying in genetics of complex diseases will be weak. We need to get used to this, as do policymakers. The example of breast cancer genes is an extraordinary, rare example: most will be small effects.

Dodge: The breast cancer genes are actually a small effect. They affect around 20% of all breast cancers. But the public is very quick to respond. A third of women in their 30s who screen positively have preventive mastectomies, which is a dramatic behavioral response.

Martinez: For that individual it is a strong effect. But for the population, the effect will be quite weak because the mutations you talk about are very rare.

Reeve: I am a molecular geneticist who has been working with DNA for many years, and I feel like I am on Mars when I hear you talk about genes. We need to get our act together before we start communicating to the public. I read press releases from some of you, and I find them quite difficult to understand. When Nick Martin said something like 'a gene is scaleable under stress', I didn't know what he was talking about. I think he means that the transcriptional apparatus is now activating that gene.

Martin: These are different ways of describing the same phenomenon.

Braithwaite: It's a language difficulty. I'd like to add that what has been obvious throughout our discussions here is that we use this term E really loosely, but no one has told us what E does to G. The induction of transcription is one possibility, but we have also heard about life history having effects on behavioral outcomes: what does life history mean to a cell or an organ? There is a memory of some event that has taken place during our life in our tissues, but what does this mean? I think we need to have some discussions about what E means to G.

Rutter: Nick is experiencing a stress challenge here!

Martin: Scaling under stress is a nicer way of putting it than molecular biology speak!

Braithwaite: Memory doesn't exist in genes that are turned on or up-regulated.

Martin: How do we know? Isn't that precisely what we think methylation might be doing?

Braithwaite: So far, our discussion has been about trying to find genes, and the problems with doing so, but not about what we mean by G × E. We can identify G × E in some manner and say it exists, but we don't know what it means.

Martin: That is exactly what Harold Snieder's paper was about. If we can find ways of up-regulating genes so that the gene effect is larger, we have a better chance of seeing it above the noise.

Snieder: We improve your signal. If we do find these genes, perhaps we can bring them back to the lab and see what they do.

Braithwaite: We have had a mouse example (Marco Battaglia's paper, this volume) of a memory of some historical event. This is a possible way of exploring the issue.

Tissues might be examined from these mice to determine if, for example, there are aberrant levels of expression of physiologically relevant genes compared to controls. One could also examine, as Nick mentioned, possible methylation variations of these genes. Such work can most easily be done in the mouse, or other model organism.

Battaglia: I see your point about alternative splicing, in terms of duration. The Acetylcholinesterase splice variants I was mentioning before are quite long-lasting *in vitro* following acute stress. But interestingly, the incretion of cortisol further facilitates the production of the same splice variant Moreover, corticosterone and forced swimming provoke long-lasting (weeks) shift from AchE-S to AchE-R splice variant and induce AchE-R mRNA translocation into the neurites. This is not to say that this is the mechanism, but it suggests that it may be one of the first effectors of the cascade.

Tesson: It would be interesting to identify the factors that are promoting this alternative splicing. Perhaps they can act on different genes. The general relationship between the gene and the phenotype, even in monogenic disorders, is not that easy to find. The genes that have been found are most of the time not in the pathophysiological pathways that were in the beginning thought to be responsible for that disease. Even if you transfect these mutations into animal models, you can't be sure that you'll find the same phenotype as in humans. This is a general problem for linking the effect of the variation and the phenotype.

Kotb: That is because this is a complex trait. This is a message we need to make clear.

Dodge: Aren't all the phenomena in which we are interested complex? Isn't it a terrible over-simplification to say that there will be a gene for, for example, alcoholism, or even a cluster of genes? Genes don't cause alcoholism directly. They operate and act in much narrower ways that we haven't discovered yet. If you take seriously the G × E interaction effect, and if we recognize that we can't quantify these but have to look at every one together, I am surprised that no one has proposed that we should measure the environment more precisely and figure out exactly what genes do. It is as absurd to say that genes cause alcoholism as it is to say that we have a biological ability to blow perfect spheres of bubblegum. What does the gene do in this analogy? It allows the individual to blow air indiscriminately. It is the environment—and physics—that determines that it will be a sphere. It is the properties of bubblegum in the atmosphere that leads one to conclude the maximum volume can be contained in a sphere. Genes do something extremely important; environments also do something important, and it is the combination that leads to alcoholism or conduct disorder or other psychiatric conditions.

Kotb: Let's consider a hypothetical scenario where a finding indicates that certain complex genetic make-up is associated with certain behaviors or characteristics.

What is the ethics of reporting or not reporting this? If it is politically incorrect to report this finding, does the scientist suppress it?

Dodge: I wouldn't advocate not reporting it, but I would be very careful about the manner in which it is reported. I think it is flawed to hide behind the notion that one should call it as one sees it, and there is no further responsibility on the part of the scientist.

Martinez: This dilemma is a real one. I live in Arizona and have a clinic in the Apache reservation. It is clear that the Apaches have a series of diseases that are genetic, as the result of inbreeding which stems from the way they were treated. They refuse to have any genetic study done in their population. They don't even provide us with DNA because they fear it will be used to discriminate against them. Their tribe has forbidden any kind of genetic study for the foreseeable future.

Rutter: Ken Dodge, you were saying that you think it is rare to have traits that have good effects in some people but bad effects in others. You then went on to say that these have been looked at very little. In your own field, high anxiety puts you substantially at risk for anxiety disorders but it is also protective against antisocial behavior. There are genetic effects that make you more vulnerable to genetic diseases of various kinds, but also make those diseases more serious once you have them. Evolutionary theory would argue that it is very unlikely that there are genes that lead to increased or decreased vulnerability to adversity or stress. It is much more likely that there are genes that influence responsivity to the environment, and that this may be good or bad in different circumstances. Again, the example of so-called 'difficult' temperament is interesting: in starvation conditions in Africa it was protective (De Vries 1984), but in most circumstances this temperamental feature is associated with psychopathological risk (Rutter 1989).

Although it would be very hard to have a strong policy message, it does indicate that we are dealing with dynamic systems in which (with multifactorial disorders) there is not a simple subdivision into bad genes that you should get rid of and good genes that you should promote. The mix will be advantageous for some outcomes and risky for others. The future surely lies in understanding much better than we do the mediating mechanisms. Our solid findings on G × E interaction are still relatively limited (although they are growing), but they provide a way in to the causal pathways for both the genes and the environment. This won't instantly tell us what to do, but in the longer term it is likely to do so.

Dodge: There is not a good or bad individual, group or gene. Genes have survived because of their functional value, and sometimes they meet upon environments in which they are no longer functional. One might make a general hypothesis that many disorders are adaptive functional responses gone slightly awry. As an example, aggression in response to provocation and threat is probably healthy, but when it exceeds a limit or is in the wrong context, it is a problem. But the public and poli-cymakers think in terms of good genes and bad genes, and some of the research we do and the way we report it contributes to this misconception.

Rutter: With environmental influences, as with genetic factors, there is a need to consider a range of possible outcomes. For example, there are many observa-tional studies suggesting that light to moderate alcohol consumption reduces the risk of heart disease (see O'Keefe et al 2007). Although the apparent cardioprotec-tive effects might be a function of unmeasured confounders, the evidence that alcohol increases high density lipoproteins (and other protective features) in a dose-response way points to the likelihood that the benefits may be real. On the other hand, the overall risk/benefit ratio needs also to take into account the ten-dency for moderate drinking to slide into heavy binge drinking with all the serious problems associated with liver damage, motor vehicle accidents and other major sequelae. The simple dichotomous division into either wholly good or wholly bad genes or environment is likely to be misleading.

Martin: It occurred to me, that if we look at the Wellcome Trust paper (Wellcome Trust Case Control Consortium 2007) with the spectacular genome-wide associa-tion scan for several diseases, the outstanding success is type 2 diabetes. Is it just coincidental that this is a disease of modern living, where we have seen a transition from a hunter–gatherer to an oversupply of calories, and you have increased the stress? You have magnified the gene effects for this disease, where multiple SNPs have been found, but no SNPs have been found for bipolar disease or hypertension where arguably the environment hasn't had such a dramatic change.

References

Belsky J, Bakernmans-Kranenburg MJ, van Ijzendoorn MH 2007 For better and for worse: differential susceptibility to environmental influences. Curr Dir Psychol Sci 16:300–304

De Vries MW 1984 Temperament and infant mortality among the Masai of East Africa. Am J Psychiatry 141:1189–1194

Flegal KM, Graubard BI, Williamson DF, Gail MH 2007 Cause-specific excess deaths associ-ated with underweight, overweight, and obesity. JAMA 298:2028–2037

O'Keefe JH, Bybee KA, Lavie CJ 2007 Alcohol and cardiovascular health: the razor-sharp double-edged sword. J Am Coll Cardiol 50:1009–1014

Rutter M 1989 Temperament: conceptual issues and clinical implications. In: Kohnstamm GA, Bates JE, Rothbart MK (eds) Temperament in childhood. Wiley, Chichester, p 463–479

Wellcome Trust Case Control Consortium 2007 Genome-wide association study of 14,000 cases of seven common diseases and 3,000 shared controls. Nature 447:661–678

7. Gene–environment interaction and the metabolic syndrome

Kristi B. Adamo and Frédérique Tesson*

*Healthy Active Living and Obesity (HALO) Research Group, Chalmers Research Group, Children's Hospital of Eastern Ontario (CHEO) Research Institute, Ottawa, Canada and * University of Ottawa Heart Institute, Ottawa, Ontario, Canada*

Abstract. The metabolic syndrome, which has been shown to affect as many as 20% of the general adult US population, is generally described as a cluster of cardiovascular risks factors, most notably obesity, type 2 diabetes or resistance to insulin-stimulated glucose uptake (insulin resistance), dyslipidaemia and hypertension. All these risk factors are under both genetic and environmental control; they are considered individually as complex genetic diseases. Prior to pharmacological interventions for hypertension, diabetes and dyslipidaemia, lifestyle changes, in particular weight loss (or weight maintenance) and physical activity, were prioritized and constituted an effective first-line intervention strategy. Here we want to focus on three clinical components of the metabolic syndrome and the environmental factors that are considered to be the most significant targets for primary interventions: type 2 diabetes and exercise, obesity and diet, and hypertension and salt. Our experimental approach is to go from candidate gene strategy to genome-wide association. The identification of the genetic component of these risk factors is a major challenge, and it is hoped that this would help unravel mechanistic pathways that can ultimately serve as new targets for therapeutic intervention.

2008 Genetic effects on environmental vulnerability to disease. Wiley, Chichester (Novartis Foundation Symposium) p 103–121

Even though its existence has been challenged on a regular basis, the metabolic syndrome is generally described as a cluster of cardiovascular risks factors, most notably obesity, type 2 diabetes or resistance to insulin-stimulated glucose uptake (insulin resistance), dyslipidaemia, and hypertension. This aggregation of multiple risk factors has been shown to affect as many as 20% of the general adult US population (Park et al 2003) and approximately 44% of the US population over 50 years of age (Alexander et al 2003). A number of clinical studies have confirmed that the metabolic syndrome is a strong predictor of cardiovascular diseases, and, in non-diabetic people, of the development of type 2 diabetes. While there is no question that certain cardiovascular risk factors are prone to cluster, the metabolic syndrome's value as cardiovascular risk marker has been questioned. This might be related to the imprecise definitions used as diagnostic tools. One of the most recent definitions of the metabolic syndrome, also called the insulin resistance syndrome or the syndrome X, was provided by the International Diabetes Federation (Alberti et al 2005). To be diagnosed with the metabolic syndrome, a individual should display central obesity, assessed by waist circumference, plus any two of four additional criteria: raised triglyceride levels ($\geq$1.7 mmol/l), reduced high density lipoprotein (HDL) cholesterol (<1.03 mmol/l in males, <1.29 mmol/l in females), elevated blood pressure (systolic blood pressure $\geq$ 130 mmHg or diastolic blood pressure $\geq$ 85 mmHg), raised fasting plasma glucose ($\geq$5.6 mmol/l), or be receiving a specific medical treatment for any of these four conditions (Alberti et al 2005). Most of these risk factors are age, gender and ethnicity dependent, and notably each of them is controlled by both genetic and environment factors resulting in variability in expression. Lifestyle changes, in particular weight loss (or weight maintenance) and physical activity, deserve priority and constitute an effective first-line intervention strategy. Then, pharmacological interventions for hypertension, diabetes and dyslipidaemia are considered, and finally surgical therapies for selected obese patients.

Here we will focus on three clinical components of the metabolic syndrome and the most significant environmental factors that are believed to contribute to their onset and are also considered as the primary intervention strategies: type 2 diabetes and exercise, obesity and diet, and hypertension and salt. Our experimental approach is to go from candidate gene strategy to genome wide association (GWA).

The peroxisome proliferator activated receptor γ (PPARγ) and the metabolic syndrome

Our first two examples are based on candidate gene strategy and focused on PPARγ pathways. PPARγ is considered as a strong, if not the strongest, candidate gene for the metabolic syndrome. The PPARγ gene is located at 3p25, a region showing evidence for linkage with diabetes and obesity susceptibility. Frameshift and missense heterozygous mutations have been liked to insulin resistance and type 2 diabetes, obesity, lipodystrophy and hypertension (Ristow et al 1998, Barroso et al 1999, Hegele et al 2002, Savage et al 2002). PPARγ is a fatty acid- and eicosanoid-dependent nuclear receptor that binds to specific DNA response elements (PPREs) as heterodimer with the retinoid X receptor and, in the presence of ligands, regulates the expression of the target gene. Although the role of PPARγ in adipose tissue development and function is established, its low levels in tissues important to glucose homeostasis, including skeletal muscle, liver, and pancreatic β cells, raise the question of its possible physiological and pharmacological importance at those sites (Semple et al 2006). At the skeletal muscle level in particular, the total mass of muscle and its function as the site of 70% of insulin-mediated glucose disposal suggest physiologically important effects of PPARγ (Semple et al 2006). Furthermore, synthetic PPARγ agonists, the insulin-sensitizing thiazolidinediones (TZDs), are therapeutic agents used in the treatment of type 2 diabetes. However, clinical use of TZDs is limited by the occurrence of fluid retention, haemodilution, and heart failure in up to 15% of the patients (Mudaliar et al 2003).

By far the most studied PPARγ polymorphism is the Pro12Ala in the unique PPARγ2 N-terminal domain. *In vitro*, the Pro12Ala polymorphism exhibits reduced binding to DNA and modest impairment of transcriptional activation (Deeb et al 1998). Discordant results have been reported for the association of the Pro12Ala polymorphism with diabetes. However, a meta-analyses demonstrated an association of Pro12Ala with type 2 diabetes (Altshuler et al 2000). It is possible that the environment, particularly exercise and/or diet habits, could modify the genotype effect and may account for the disparity found between separate studies.

Gene–exercise interaction in type 2 diabetes

When studying gene–environment interaction on the quantitative traits that underlie diabetes, the power to detect interaction is highly dependent on the precision with which non-genetic exposures are measured (Wareham et al 2002). Achievement of optimal glycaemic control is the focus of traditional treatment paradigms. Regular exercise, both aerobic (walking, jogging, or cycling) and resistance (weightlifting) training results in increased glucose uptake and insulin sensitivity and is a primary modality used in the treatment of type 2 diabetes patients (Sigal et al 2007). Similar to pharmaceutical treatment, large inter-individual exercise training responses are observed and are hypothesized to be dependent on genetic background. We addressed the possible PPARγ2 Pro[12]Ala genotype-glucose response interaction prospectively, using a closely monitored exercise intervention in a population of type 2 diabetes patients (Adamo et al 2005).

The 139 type 2 diabetic subjects included in the present genetic study completed 3 months of supervised exercise training intervention as part of the Diabetes Aerobic & Resistance Exercise (DARE) study (Sigal et al 2007). The present study did not include a specific dietary intervention. The study cohort includes previously sedentary Caucasian patients aged 40–70 years with type 2 diabetes, with or without oral agents but not treated with insulin, with HbA1c between 6.6 and 9.9 %. The conduct of the clinical trial was described as 'exemplary' in the editorial commentary of the paper and demonstrates the direct, clinical relevance of exercise training for the management of glucose levels (Kraus & Levine 2007). The exercise training intervention led to similar improvements in body composition and glucose homeostasis variables in both genotype groups ($P < 0.05$). The change in fasting plasma glucose was significantly different between PPARγ2 genotypes (−1.66 mmol/l vs. −0.54 mmol/l, Ala carriers and wildtype respectively) ($P =$ 0.034 unadjusted and $P =$ 0.089 including baseline glucose) and the significant association between genotype and glucose response remained after adjusting for statistically significant predictors (age, changes in insulin and BMI [$P =$ 0.015]) and including the baseline glucose, insulin and BMI ($P =$ 0.031).

Exercise training and the Ala allele must act either independently or in synergy to modify glucose homeostasis through increasing glucose uptake or by decreasing hepatic glucose output. At the whole body level, exercise training has been shown to increase insulin sensitivity (Borghouts & Keizer 2000, Short et al 2003, Duncan et al 2003) and has also been shown to decrease basal hepatic glucose production in patients with type 2 diabetes (Segal et al 1991). The Ala allele has been associated with more efficient insulin suppression of glucose production (Muller et al 2003) as well as greater insulin clearance (Tschritter et al 2003). It has been proposed that carriers of the Ala allele have greater insulin sensitivity following embarking upon a physical activity regime (Weiss et al 2005, Kahara et al 2003) and that the Pro allele carriers were more likely to have type 2 diabetes if they were physically inactive (Nelson et al 2007). The findings in our population of type 2 diabetes patients support this hypothesis as Ala carriers showed an improvement in fasting plasma glucose despite similar changes in fasting insulin. Physical activity has been suggested to modify the risk of developing type 2 diabetes associated with other SNPs in genes regulating insulin secretion such as the genes encoding GLUT2 and SUR1 (Kilpelainen et al 2007). In summary, the present study conducted using a well-controlled, individualized exercise intervention, suggests that the PPARγ2 Pro12Ala SNP influences the fasting glucose response to exercise training in previously sedentary type 2 diabetic patients. These results, which should be confirmed by other independent studies, suggest a possible benefit for tailored genotype-specific treatment interventions. Ala carriers may be particularly susceptible to the negative metabolic consequences of sedentary living, and they conversely would have most to gain from a targeted preventive exercise training intervention.

Gene–diet interaction in obesity

The current management of obesity is primarily to reduce energy intake and increase energy expenditure. However, many people are resistant to weight loss, or rapidly regain lost weight after cessation of treatment. Identifying individuals responsive to lifestyle modification to control body weight would be incredibly advantageous.

Performing a genome scan is the only unbiased strategy for identifying genes related to weight loss in response to treatment intervention. To date, there is no relevant genome scan for body-composition phenotypes in response to diet. A recent study shows that human populations that have high starch diets have an increase in the number of copies of the salivary amylase gene (AMY1) whose product breaks down starch (Perry et al 2007). Although copy number variation (CNV) has attracted a lot of recent attention, this is one of the first documented examples of positive selection on CNV in humans. CNVs are common gains or losses of large chunks of DNA sequence which might alter gene dosage, disrupt coding sequences, and perturb long-range gene regulation. The majority of studies, including the one described in details here, were generated from small-scale candidate gene association studies commonly accompanied by lack of replication and statistical pitfalls (Table 1) (Adamo & Tesson 2007). Moreover, the functional significance of the associated polymorphism is very often not known and thus the link between genetic and molecular mechanism can only been speculated.

We focused on determining whether both PPARγ and acyl CoA synthetase 5 (ACSL5) genes, the ACSL5 being one of the putative PPARγ response genes, were involved in the inter-individual response to dietary treatment (Adamo et al 2007). The PPARγ response gene acyl CoA synthetases catalyze the activation of

TABLE 1   Polymorphisms associated with body-composition phenotypes in response to diet only

Gene polymorphism	Population	Intervention	Conclusions	Reference
LEP (Leptin) C-2549A	n = 79 Obese women	Low-calorie diet (25% reduction in energy intake)	C allele associated with a greater change in BMI	Mammes et al 2001
LEPR (Leptin Receptor) Ser343Ser (T/C)	n = 116 Overweight women	Low-calorie diet (25% reduction in energy intake)	C allele associated with more weight loss	Mammes et al 2001
5HT$_{2C}$ (Serotonin Receptor 2C) Cys23Ser	n = 148 Teenage girls	Weight loss over time in teenage years	Ser substitution associated with weight loss	Westberg et al 2002
UCP1 (Uncoupling protein 1) A-3826G	n = 163 Overweight individuals	10 week low calorie diet	G/G carriers more resistant to weight change	Fumeron et al 1996
APOA5 (Apolipoprotein A5) T-1131C	n = 606 Hyperlipemic overweight men	Short-term, monitored fat restriction diet	C carriers lost more weight than T homozygotes	Aberle et al 2005

(*Continued*)

TABLE 1 *Continued*

Gene polymorphism	Population	Intervention	Conclusions	Reference
PLIN (Perilipin) G11482A (intronic)	*n* = 48 Obese individuals	2 week very low energy diet; 603 kcal/d followed by 1200 kcal/d for 1 year	GG genotype predictive of weight loss	Corella et al 2005
Adipsin Hinc II RFLP	*n* = 12 Twin pairs	100 days of overfeeding	Absence of HincII RFLP associated with greater weight gain	Ukkola et al 2001
GRL (Glucocorticoid Receptor) C/G intron 2	*n* = 120 Obese individuals	6 week diet + 1 year follow-up	G/G genotype predictor of weight maintenance	Vogels et al 2005
	n = 12 Twin pairs	100 days of overfeeding	C/C genotype associated with greater weight gain and abdominal visceral fat	Ukkola et al 2001
ACSL5 (Acyl CoA Synthetase 5) rs2419621C/T (5' flanking region)	*n* = 141 Obese women	6 week caloric restriction	T allele associated with greater response	Adamo et al 2007
PPARγ2 (Peroxisome Proliferator Activated Receptor γ2)	*n* = 70 Overweight/ Obese postmenopausal women	6 month hypocaloric diet	Ala allele associated with weight regain	Nicklas et al 2001
Pro12Ala	*n* = 141 Obese women	6 week caloric restriction	Ala allele associated with resistance to weight loss	Adamo et al 2007
ADRB3 (β3- Adrenergic Receptor) Trp64Arg	*n* = 24 Obese women	Medically supervised 1200 kcal/ day diet	Arg allele associated with resistance to loss of visceral fat	Tchernof 2000
ADRB2 (β2- Adrenergic Receptor) Gln27Glu Arg16Gly	*n* = 12 Twin pairs	100 days of overfeeding	Gln/Gln genotype associated with greater weight gain response	Ukkola et al 2001

intracellular long-chain fatty acids into long-chain fatty acyl-CoA that are sequentially used for lipid synthesis, protein modification, and β-oxidation. The ACSL5 gene maps to the obesity locus 10q25.1-2 (van der Kallen et al 2000), we hypothesized that ACSL5 genotype may influence the rate of weight loss in response to energy restriction.

Genotypic/phenotypic comparisons were made between selected obese women from the quintiles losing the most (diet responsive, $n = 74$) and the quintiles losing the least weight (diet-resistant, $n = 67$) in the first six weeks of a 900 kcal formula diet (Harper et al 2002). Two common PPARγ single nucleotide polymorphisms (SNP), Pro12Ala and C1431T, and eight polymorphisms across the ACSL5 gene, were selected for single locus and haplotypic association analyses. Although it is possible that the causal genetic variant is included in the set of variants/markers tested, the premise for this approach is that one or more of the variants tested will serve as a proxy for the causal variant. In this study, the number of polymorphisms analyzed was sufficient to assume that the full extent of nucleotide sequence variation was captured. The PPARγ Pro^{12}Ala SNP was associated with diet resistance (odds ratio = 3.48, 95% CI = 1.41–8.56, $P = 0.03$) and the rs2419621, located in the 5'UTR of the ACSL5 gene, displayed the strongest association with diet response (odds ratio = 3.45, 95% CI = 1.61–7.69, $P = 0.001$). Although we found modest evidence for an interaction between these two loci ($P = 0.058$), studies with a greater number of subjects are required to confirm this relationship. In addition, skeletal muscle ACSL5 mRNA expression was significantly lower in carriers of the wild-type as compared to the variant rs2419621 allele ($P = 0.03$). Further studies showed that the rs2419621 polymorphism is functional and able to modulate the expression levels in vitro (Tesson et al, manuscript in preparation). Our results suggest a link between PPARγ2 and ACSL5 genotype, and diet responsiveness.

Gene–salt interaction in hypertension

It is believed that in the general population about 10% of individuals are salt-sensitive meaning that high salt intake increases the blood pressure in the short-term. However, as many as 50% of hypertensive patients are salt-sensitive in short-term studies. There is a strong heritability component. It has been shown that the correlation coefficient for difference in blood pressure between regular vs. low salt intake can reach 0.7 for identical twins (Miller et al 1987). Ethnicity is a major determinant of salt-sensitivity. One study showed that salt-sensitivity was present in as many as 56% of black individuals compared to only 29% of whites (Wilson et al 1999). The prevalence of salt sensitivity increases in older age groups (de la Sierra et al 1995). It is believed that estrogens blunt the effect of high salt on blood pressure and indeed, surgical menopause has been shown to increase the salt sensitivity of blood pressure (Schulman et al 2006).

To date, there is no GWA study for salt-sensitive hypertension. Based on animal models, several associations have been studied with genes from the RAAS, from the sympathetic nervous system, and from other pathways (Table 2). The few studies that assessed the simultaneous effect of several candidate genes on salt sensitivity of blood pressure did not find any synergistic effect. In summary, so far very little is known about the genes that are responsible for the salt sensitivity of blood pressure. We can consider two broad classes of pitfalls: genotyping and phenotyping pitfalls.

TABLE 2   Polymorphisms associated with salt-sensitive hypertension

Gene/polymorphism	Population	Assessment salt-sensitivity/blood pressure	Conclusions	Reference
ACE/ID	n = 2823 Normotensive Japanese	Questionnaires & 24 h urine collection/ office measurement	High sodium intake strengthens the association of ACE with blood pressure	Yamagishi et al 2007a
CYP11B2/IC	n = 2823 Normotensive Japanese	Questionnaires & 24 h urine collection/ office measurement	No association	Yamagishi et al 2007b
AGTR2/C3123A	n = 217 Japanese men	Questionnaires/ office measurement	AGTR2 A allele associated with salt-sensitivity	Miyaki et al 2006
ACE/ID AGTR1/A1166C & C-344T CYP11B2/IC GNB3/C825T	n = 102 Hypertensive Caucasians	Weinberger protocol/office measurement	CYP11B2/IC allele associated with salt-sensitivity	Pamies-Andreu et al 2003
ACE/ID HSD11B2/G534A	n = 71 Hypertensive Caucasians	Dietary protocol/ casual & AMBP	ACE I & HSD11B2 G alleles associated with salt-sensitivity	Poch et al 2001
ACE/ID AGTR1/A1166C AGT/M235T	n = 50 Hypertensive Caucasians	Dietary protocol/AMBP	ACE I allele associated with salt-sensitivity	Giner et al 2000

Gene/polymorphism	Population	Assessment salt-sensitivity/blood pressure	Conclusions	Reference
ACE/ID AGT/M235T	n = 46 Caucasians	Increasing salt intake/AMBP	AGT M allele associated with salt-sensitivity	Johnson et al 2001
AGT/M235T	n = 86 Untreated Hypertensives	Dietary sodium reduction/office measurement	AGT T allele associated with decrease in blood pressure	Hunt et al 1999
HSD11B2 intron 1 CA repeat	n = 198 Caucasians	Weinberger protocol/office measurements	Small number of CA repeat associated with salt-sensitivity	Agarwal et al 2000
HSD11B2 AluI & microsatellite	n = 149 Caucasian males	Dietary protocol/ office measurements	HSD11B2 microsatellite associated with salt-sensitivity	Lovati et al 1999
ADRB2 G46A & C79G	n = 171 Hypertensive & 65 normotensive Caucasians	Low & high salt-diet/office measurements	A46A/C79C diplotype associated with salt-sensitivity	Pojoga et al 2006
GNB3/C825T	n = 193 normotensive Caucasians	Low & high salt-diet/office measurements	No association	Schorr et al 2000
	n = 46 Caucasians	Low & high salt-diet/AMBP	No association	Gonzalez-Nunez et al 2001
	n = 78 Hypertensive & 76 normotensive Caucasians	Weinberger protocol/office measurements	No association	Martin et al 2005
GRK4/R65L, A142V, V247I, A253T, A486V & G562D AGT/M235T AGTR1/A1166C CYP11B2/C-344T PAI1/4/5G GNB3/C825T HSD11B2/G534A	n = 184 Hypertensive Japanese	Low & high salt-diet/office measurements & AMBP	GRK4 R65L, A142V & A486V associated with salt-sensitivity	Sanada et al 2006

(*Continued*)

TABLE 1   *Continued*

Gene/polymorphism	Population	Assessment salt-sensitivity/blood pressure	Conclusions	Reference
ENOS/T-786C E298D	n = 281 Normotensive Japanese	Questionnaires/ office measurements	ENOS C allele associated with salt-sensitivity	Miyaki et al 2005
	n = 126 Normotensive Hispanics	Low & high salt-diet/office measurements	ENOS 4a/b associated with salt-sensitivity	Hoffmann et al 2005
ADM/A-1984G	n = 427 Chinese	Urine collection/ office measurement & AMBP	ADM G allele associated with lower sodium excretion	Li et al 2006
ADD1/G460W	n = 379 Caucasians	Urine collection/ office measurement	ADD1 W allele associated with plasma ouabain	Wang et al 2003
	n = 126 Hispanics	Low & high salt-diet/office measurements	No association	Castejon et al 2003
	n = 117 Normotensive Caucasians	Dietary protocol/ office measurements	ADD1 WW associated with salt sensitivity	Beeks et al 2004
	n = 86 Hypertensives	Weinberger protocol/office measurements	ADD1 G allele associated with decreased blood pressure	Cusi et al 1997
	n = 123 Hypertensive Caucasians	Weinberger protocol/casual & ABPM	ADD1 WW associated with salt sensitivity	Manunta et al 1998
	n = 279	Low & high salt-diet/office measurements	ADD1 WW associated with salt sensitivity	Grant et al 2002
CLCNKA/4SNPs	n = 314 Never treated Caucasians	300 mM i.v. saline /ABPM	Haplotype of 4 SNPs associated with salt-sensitivity	Barlassina et al 2007

By nature the candidate gene strategy is biased since it is hypothesis driven. It depends on the correct *a priori* understanding of the actual primary pathways that initiate the disease. Unfortunately, this is far from clear for most complex clinical disorders. Those studies focus on a small number of gene variants while disregarding other genes within the same pathway and overlooking potential gene–gene interactions. If genes act together, the marginal effect of each gene might be small, but collections of genes might have much larger effects. Most of the studies are underpowered, due to inadequate sample size. The last point is the lack of reproducibility; the potential reasons for this include false-positive associations which are correctly not replicated, or false negative associations which is the case when a true association is not replicated in an underpowered follow-up study, and finally a true association in one population may not be true in a second population because of heterogeneity in genetic or ethnic background (Newton-Cheh & Hirschhorn 2005), such as the difference in blood pressure response to high salt intake between blacks and whites. Another drawback is the lack of stratification according to gender and age.

The phenotyping pitfalls fall into three categories (Tesson & Leenen 2007). First, in the vast majority of studies, the blood pressure was measured by office measurements. These are notoriously variable and inaccurate and can lead to over-diagnose the 'white coat hypertension' or under-diagnose the 'masked hypertension'. Very few studies employed the current gold standard, the 24 hour ambulatory blood pressure monitoring. On top of casual assessment of blood pressure, the sodium intake is usually measured from a single urine sample or even worse by using questionnaires on average food intake. Lastly, in most studies, patients are being treated with antihypertensive medication—an obvious confounding factor.

Determining which measures to apply is a trade-off between sample size and the validity of the assessment. However, the statistical power of interaction studies shows a steep drop when environmental exposures and phenotypic outcomes are inaccurately measured (Wong et al 2003). To overcome previous limitations, in collaboration with Dr Frans Leenen from the University of Ottawa Heart Institute, we embarked on a GWA study in a population of Caucasian subjects with early onset hypertension and a positive family history for hypertension in order to enrich for risk alleles, carefully phenotyped for blood pressure response to salt on 24 h ambulatory blood pressure monitoring. All measurements are completed after stopping the antihypertensive drug therapy for 2 weeks. The salt sensitivity is measured by maintaining a low salt diet (~100 mmol/day) for two three-week periods, then by administrating a sodium supplement of 200 mmol/day to reach about 300 mmol/day in total for one three-week period. At the end of each three-week period, urine is collected for 24 hours and the blood pressure is monitored for 24 hours. GWA studies, the brute-force forms of discovery science (Loscalzo

2007), are the only unbiased assessment of gene–environment interaction because they are model and hypothesis-free. Such an approach is the only way to identify previously unrecognized chromosomal loci but even more importantly the multiple influencing genes. We will use the Affymetrix® Genome-Wide Human SNP Array 6.0, which features 1.8 million genetic markers, including more than 900 000 SNPs and more than 940 000 probes for the detection of CNVs (copy number variants). Interrogation of the genome for both types of variants may be an effective way to elucidate the causes of complex diseases and allow to ultimately translating genetic variants into personalized genetic medicine.

From a public health perspective, high salt intake has a clear effect on blood pressure and thereby cardiovascular morbidity and mortality. From an individual perspective, the impact of high salt intake on his/her cardiovascular system can vary from minimal to substantial and appears to a large extent genetically determined. The extent of this impact is clinically difficult to ascertain. Genetic diagnosis would be an ideal method of choice for advising lifestyle interventions for a particular individual.

Perspectives

The scientific community investigating gene–environment interactions for the metabolic syndrome is faced with three major challenges: the need for (1) large sample size to ensure sufficient statistical power in association studies; (2) detailed phenotypic and environmental characterization of participants; and (3) sophisticated statistical methods to manage the large amount of generated comparisons. Lack of statistical power, inaccuracy in the assessment of outcome phenotypes and environmental exposure contributes to the difficulties in finding and replicating interactions (Grarup & Andersen 2007). The next wave of interrogation of the genome using GWAs may be an effective way to elucidate gene–environment interactions and allow to ultimately translating genetic variants into personalized genetic medicine.

Nevertheless, statistical evidence alone cannot distinguish between causal and non-functional variants in linkage disequilibrium with the causal polymorphism. Therefore, the evidence for causality must include compelling demonstration that variations can be translated into a phenotype to achieve significant relevance to diagnostics and therapeutics. Despite ongoing debates over the power of genetics to estimate common disease risk, it is widely believed that genetic information will become increasingly important in healthcare.

However, in addition to genetic information, epigenetic alterations can no longer be ignored in evaluations of the causes of the complex disorders. There are many covalent epigenetic modifications that keep genes stably repressed or active. The best-studied epigenetic modification is DNA methylation which, in most cases, induces the silencing of gene expression. Histone post-transcriptional modifications, including acetylation, methylation, ubiquitination and phosphorylation, can also modulate chromatin structure, accessibility to transcription factors and, thus, gene expression. A growing body of evidence suggests that small interfering RNA (siRNA)-mediated mechanisms play a central role in regulating gene activity. Individuals with metabolic syndrome may have suffered improper epigenetic programming during their early development due to placental insufficiency, inadequate maternal nutrition and metabolic disturbances (Junien & Nathanielsz 2007). Epigenetic misprograming during development is widely thought to have a persistent effect on the health of the offspring and may even be transmitted to the next generation (Junien & Nathanielsz 2007). A recent work showed a direct link between maternal low protein diet and undermethylation of the AT1b angiotensin receptor gene promoter in the offspring correlated with an increased expression of the AT1b angiotensin receptor and the development of hypertension in a rat model, thus supporting the Barker hypothesis (Bogdarina et al 2007). Moreover, during ageing, an epigenetic drift is observed, consisting in a global change in methylation of CpG in several gene promoters thus altering their level of expression (Junien & Nathanielsz 2007). At this point, there is a need for systematic large-scale epigenetic studies to understand how the epigenetic control of the genome contributes to diseases.

Acknowledgments

Grants: Frederique Tesson, Operating Grants from the Canadian Institutes of Health Research (CIHR MOP 77685), and from the Heart and Stroke Foundation of Ontario (HSFO # NA 5849).

References

Aberle J, Evans D, Beil F, Seedorf U 2005 A polymorphism in the apolipoprotein A5 gene is associated with weight loss after short-term diet. Clin Genet 68:152–154

Adamo KB, Tesson F 2007 Genotype-specific weight loss treatment advice: how close are we? Appl Physiol Nutr Metab 32:351–366

Adamo K, Sigal R, Williams K, Kenny G, Prud'homme D, Tesson F 2005 Influence of Pro 12 Ala peroxisome proliferator-activated receptor γ2 polymorphism on glucose response to exercise training in type 2 diabetes. Diabetologia 48:1503–1509

Adamo K, Dent R, Langefeld C et al 2007 Peroxisome proliferator-activated receptor γ2 and acyl-CoA synthetase 5 polymorphisms influence diet response. Obes Res 15:1068

Agarwal A, Giacchetti G, Lavery G et al 2000 CA-repeat polymorphism in intron 1 of HSD11B2: effects on gene expression and salt sensitivity. Hypertension 36:187–194

Alberti K, Zimmet P, Shaw J 2005 The metabolic syndrome—a new worldwide definition. Lancet 366:1059–1062

Alexander C, Landsman P, Teutsch S, Haffner S 2003 NCEP metabolic syndrome, diabetes mellitus and prevalence of coronary heart disease. Diabetes 52:1210–1214

Altshuler D, Hirschhorn J, Klannemark M et al 2000 The common PPAR-gamma Pro12Ala polymorphism is associated with decreased risk of type 2 diabetes. Nat Genet 26:76–80

Barlassina C, Dal Fiume C, Lanzani C et al 2007 Common genetic variants and haplotypes in renal CLCNKA gene are associated to salt-sensitive hypertension. Hum Mol Genet 16:1630–1638

Barroso I, Gurnell M, Crowley V et al 1999 Dominant negative mutations in human PPAR-gamma associated with severe insulin resistance, diabetes mellitus and hypertension. Nature 402:880–883

Beeks E, van der Klauw M, Kroon A, Spiering W, Fuss-Lejeune M, de Leeuw P 2004 α-Adducin Gly460Trp polymorphism and renal hemodynamics in essential hypertension. Hypertension 44:419–423

Bogdarina I, Welham S, King P, Burns S, Clark A 2007 Epigenetic modification of the renin-angiotensin system in the fetal programming of hypertension. Circ Res 100:520

Borghouts L, Keizer H 2000 Exercise and insulin sensitivity: a review. Int J Sports Med 21:1–12

Castejon A, Alfieri A, Hoffmann I, Rathinavelu A, Cubeddu L 2003 Alpha-adducin polymorphism, salt sensitivity, nitric oxide excretion, and cardiovascular risk factors in normotensive hispanics. Am J Hypertens 16:1018–1024

Corella D, Qi L, Sorli J et al 2005 Obese subjects carrying the 11482G > A polymorphism at the perilipin locus are resistant to weight loss after dietary energy restriction. J Clin Endocrinol Metab 90:5121–5126

Cusi D, Barlassina C, Azzani T et al 1997 Polymorphisms of alpha-adducin and salt sensitivity in patients with essential hypertension. Lancet 349:1353–1357

de la Sierra A, Lluch M, Coca A, Aguilera M, Sanchez M, Sierra C, Urbano-Marquez A 1995 Assessment of salt sensitivity in essential hypertension by 24-h ambulatory blood pressure monitoring. Am J Hypertens 8:970–977

Deeb S, Fajas L, Nemoto M et al 1998 A Pro12Ala substitution in PPAR big gamma 2 associated with decreased receptor activity, lower body mass index and improved insulin sensitivity. Nat Genet 20:284–287

Duncan G, Perri M, Theriaque D, Hutson A, Eckel R, Stacpoole P 2003 Exercise training, without weight loss, increases insulin sensitivity and postheparin plasma lipase activity in previously sedentary adults. Diabetes Care 26:557–562

Fumeron F, Durack-Bown I, Betoulle D et al 1996 Polymorphisms of uncoupling protein (UCP) and beta 3 adrenoreceptor genes in obese people submitted to a low calorie diet. Int J Obes Relat Metab Disord 20:1051–1054

Giner V, Poch E, Bragulat E et al 2000 Renin-angiotensin system genetic polymorphisms and salt sensitivity in essential hypertension. Hypertension 35:512–517

González-Núñez D, Giner V, Bragulat E, Coca A, de la Sierra A, Poch E 2001 Absence of an association between the C825T polymorphism of the G-protein beta 3 subunit and salt-sensitivity in essential arterial hypertension. Nefrologia 21:355–361

Grant F, Romero J, Jeunemaitre X et al 2002 Low-renin hypertension, altered sodium homeostasis, and an α-adducin polymorphism. Hypertension 39:191–196

Grarup N, Andersen G 2007 Gene–environment interactions in the pathogenesis of type 2 diabetes and metabolism. Curr Opin Clin Nutr Metab Care 10:420–426

Harper M-E, Dent R, Monemdjou S et al 2002 Decreased mitochondrial proton leak and reduced expression of Ucp3 in skeletal muscle of obese diet-resistant women. Diabetes 51:2459–2466

Hegele R, Cao H, Frankowski C, Mathews S, Leff T 2002 PPARG F388L, a transactivation-deficient mutant, in familial partial lipodystrophy. Diabetes 51:3586–3590

Hoffmann I, Tavares-Mordwinkin R, Castejon A, Alfieri A, Cubeddu L 2005 Endothelial nitric oxide synthase polymorphism, nitric oxide production, salt sensitivity and cardiovascular risk factors in Hispanics. J Hum Hypertens 19:233–240

Hunt S, Geleijnse J, Wu L, Witteman J, Williams R, Grobbee D 1999 Enhanced blood pressure response to mild sodium reduction in subjects with the 235T variant of the angiotensinogen gene. Am J Hypertens 12:460–466

Johnson A, Nguyen T, Davis D 2001 Blood pressure is linked to salt intake and modulated by the angiotensinogen gene in normotensive and hypertensive elderly subjects. J Hypertens 19:1053–1060

Junien C, Nathanielsz P 2007 Report on the IASO Stock Conference 2006: early and lifelong environmental epigenomic programming of metabolic syndrome, obesity and type II diabetes. Obes Rev 8:487–502

Kahara T, Takamura T, Hayakawa T et al 2003 PPARγ gene polymorphism is associated with exercise-mediated changes of insulin resistance in healthy men. Metabolism 52:209–212

Kilpelainen T, Lakka T, Laaksonen D et al 2007 Physical activity modifies the effect of SNPs in the SLC2A2 (GLUT2) and ABCC8 (SUR1) genes on the risk of developing type 2 diabetes. Physiol Genomics, 31:264–272

Kraus W, Levine B 2007 Exercise training for diabetes: the "strength" of the evidence. Ann Intern Med, 147:423–424

Li Y, Staessen J, Li L et al 2006 Blood pressure and urinary sodium excretion in relation to the A-1984G adrenomedullin polymorphism in a Chinese population. Kidney Int 69:1153–1158

Loscalzo J 2007 Association studies in an era of too much information: clinical analysis of new biomarker and genetic data. Circulation 116:1866–1870

Lovati E, Ferrari P, Dick B et al 1999 Molecular basis of human salt sensitivity: the role of the 11ß-hydroxysteroid dehydrogenase type 2. J Clin Endocrinol Metab 84:3745–3749

Mammes O, Aubert R, Betoulle D et al 2001 LEPR gene polymorphisms: associations with overweight, fat mass and response to diet in women. Eur J Clin Invest 31:398–404

Manunta P, Cusi D, Barlassina C et al 1998 alpha-Adducin polymorphisms and renal sodium handling in essential hypertensive patients. Kidney Int 53:1471–1478

Martin D, Andreu E, Ramirez Lorca R et al 2005 G-protein beta-3 subunit gene C825 T polymorphism: influence on plasma sodium and potassium concentrations in essential hypertensive patients. Life Sci 77:2879–2886

Miller J, Weinberger M, Christian J, Daugherty S 1987 Familial resemblance in the blood pressure response to sodium restriction. Am J Epidemiol 126:822–830

Miyaki K, Tohyama S, Murata M et al 2005 Salt intake affects the relation between hypertension and the T-786C polymorphism in the endothelial nitric oxide synthase gene. Am J Hypertens 18:1556–1562

Miyaki K, Hara A, Araki J et al 2006 C3123A polymorphism of the angiotensin II type 2 receptor gene and salt sensitivity in healthy Japanese men. J Hum Hypertens 20:467

Mudaliar M, Face, S, Chang M, AR, Henry M, RR 2003 Thiazolidinediones, peripheral edema, and type 2 diabetes: incidence, pathophysiology, and clinical implications. Endocr Pract 9:406–416

Muller Y, Bogardus C, Beamer B, Shuldiner A, Baier L 2003 A functional variant in the peroxisome proliferator-activated receptor γ2 promoter is associated with predictors of obesity and type 2 diabetes in Pima Indians. Diabetes 52:1864–1871

Nelson T, Fingerlin T, Moss L, Barmada M, Ferrell R, Norris J 2007 Association of the peroxisome proliferator–activated receptor γ gene with type 2 diabetes mellitus varies by physical activity among non-Hispanic whites from Colorado. Metabolism 56:388–393

Newton-Cheh C, Hirschhorn JN 2005 Genetic association studies of complex traits: design and analysis issues. Mutat Res 573:54–69

Nicklas B, van Rossum E, Berman D, Ryan A, Dennis K, Shuldiner A 2001 Genetic variation in the peroxisome proliferator-activated receptor-gamma2 gene (Pro12Ala) affects metabolic responses to weight loss and subsequent weight regain. Diabetes 50:2172–2176

Pamies-Andreu E, Ramirez-Lorca R, García-Junco P et al 2003 Renin-angiotensin-aldosterone system and G-protein beta-3 subunit gene polymorphisms in salt-sensitive essential hypertension. J Hum Hypertens 17:187–191

Park Y, Zhu S, Palaniappan L, Heshka S, Carnethon M, Heymsfield S 2003 The metabolic syndrome prevalence and associated risk factor findings in the US population from the Third National Health and Nutrition Examination Survey, 1988–1994. Arch Intern Med 163:427–436

Perry GH, Dominy NJ, Claw KG et al 2007 Diet and the evolution of human amylase gene copy number variation. Nat Genet 39:1256–1260

Poch E, Gonzalez D, Giner V, Bragulat E, Coca A, de la Sierra A 2001 Molecular basis of salt sensitivity in human hypertension evaluation of renin-angiotensin-aldosterone system gene polymorphisms. Hypertension 38:1204–1209

Pojoga L, Kolatkar N, Williams J et al 2006 β-2 adrenergic receptor diplotype defines a subset of salt-sensitive hypertension. Hypertension 48:892–900

Ristow M, Muller-Wieland D, Pfeiffer A, Krone W, Kahn C 1998 Obesity associated with a mutation in a genetic regulator of adipocyte differentiation. New Engl J Med 339:953–959

Sanada H, Yatabe J, Midorikawa S et al 2006 Single-nucleotide polymorphisms for diagnosis of salt-sensitive hypertension. Clin Chem 52:352–360

Savage D, Agostini M, Barroso I et al 2002 Digenic inheritance of severe insulin resistance in a human pedigree. Nat Genet 31:379–384

Schorr U, Blaschke K, Beige J, Distler A, Sharma A 2000 G-protein beta3 subunit 825T allele and response to dietary salt in normotensive men. J Hypertens 18:855–859

Schulman I, Aranda P, Raij L, Veronesi M, Aranda F, Martin R 2006 Surgical menopause increases salt sensitivity of blood pressure. Hypertension 47:1168–1174

Segal K, Edano A, Abalos A et al 1991 Effect of exercise training on insulin sensitivity and glucose metabolism in lean, obese, and diabetic men. J Appl Physiol 71:2402–2411

Semple R, Chatterjee V, O'Rahilly S 2006 PPARγ and human metabolic disease J Clin Invest 116:581–589

Short K, Vittone J, Bigelow M et al 2003 Impact of aerobic exercise training on age-related changes in insulin sensitivity and muscle oxidative capacity. Diabetes 52:1888–1896

Sigal R, Kenny G, Boule N et al 2007 Effects of aerobic training, resistance training, or both on glycemic control in type 2 diabetes: a randomized trial. Ann Intern Med 147:357

Tesson F, Leenen FHH 2007 Still building on candidate-gene strategy in hypertension? Hypertension 50:607–608

Tchernof A 2000 Impaired capacity to lose visceral adipose tissue during weight reduction in obese postmenopausal women with the Trp64Arg beta3-adrenoceptor gene variant. Diabetes 49:1709–1713

Tschritter O, Fritsche A, Stefan N et al 2003 Increased insulin clearance in peroxisome proliferator-activated receptor γ2 Pro12Ala. Metabolism 52:778–783

Ukkola O, Tremblay A, Bouchard C 2001 Beta-2 adrenergic receptor variants are associated with subcutaneous fat accumulation in response to long-term overfeeding. Int J Obes 25:1604–1608

van der Kallen C, Cantor R, van Greevenbroek M et al 2000 Genome scan for adiposity in Dutch dyslipidemic families reveals novel quantitative trait loci for leptin, body mass index and soluble tumor necrosis factor receptor superfamily 1A. Int J Obes 24:1381–1391

Vogels N, Mariman E, Bouwman F, Kester A, Diepvens K, Westerterp-Plantenga M 2005 Relation of weight maintenance and dietary restraint to peroxisome proliferator-activated receptor γ2, glucocorticoid receptor, and ciliary neurotrophic factor polymorphisms 1 2 3. Am J Clin Nutr 82:740–746

Wang J, Staessen J, Messaggio E et al 2003 Salt, endogenous ouabain and blood pressure interactions in the general population. J Hypertens 21:1475–1481

Wareham N, Franks P, Harding A 2002 Establishing the role of gene-environment interactions in the etiology of type 2 diabetes. Endocrinol Metab Clin North Am 31:553–566

Weiss E, Kulaputana O, Ghiu I et al 2005 Endurance training–induced changes in the insulin response to oral glucose are associated with the peroxisome proliferator–activated receptor-gamma 2 Pro12Ala genotype in men but not in women. Metabolism 54:97–102

Westberg L, Bah J, Rastam M et al 2002 Association between a polymorphism of the 5-HT2C receptor and weight loss in teenage girls. Neuropsychopharmacology 26:789–793

Wilson D, Sica D, Miller S 1999 Ambulatory blood pressure nondipping status in salt-sensitive and salt-resistant black adolescents. Am J Hypertens 12:159–165

Wong M, Day N, Luan J, Chan K, Wareham N 2003 The detection of gene–environment interaction for continuous traits: should we deal with measurement error by bigger studies or better measurement? Int J Epidemiol 32:51–57

Yamagishi K, Tanigawa T, Cui R et al 2007a High sodium intake strengthens the association of ACE I/D polymorphism with blood pressure in a community. Am J Hypertens 20: 751–757

Yamagishi K, Tanigawa T, Cui R et al 2007b Aldosterone synthase gene T344C polymorphism, sodium and blood pressure in a free-living population: a community-based study. Hypertens Res 30:497–502

DISCUSSION

Chia: It seems that it is easier to translate your findings into clinical practice or policy than some of the other examples we have heard at this meeting. Perhaps in gene × environment (G × E) interaction studies we should be focusing on disease groups where environmental interventions could be applied to specific groups. However, in the public health arena it is much more difficult to study this. Perhaps our focus should be more on disease groups and looking at interventions for specific subgroups of patients, rather than the general population.

Tesson: Our results are preliminary, and should be replicated in larger groups, and independent groups before lifestyle recommendations can be made.

Chia: They may be preliminary, but it appears that the application of this work is clearer than, say, in a public health arena in the general population where you try certain kinds of interventions for certain susceptible groups.

Tesson: The environmental factors for each of the diseases discussed in my presentation are the ones that matter the most in the disease. They are primary intervention targets. Exercise is an effective intervention strategy for type 2 diabetes. The same thing applies for obesity and diet. 50% of hypertensive patients are believed to be salt sensitive. In those particular diseases, there is an obvious environmental factor which has a documented impact on the progression of the disease.

Uher: I wonder about the slightly longer-term outcomes in these interventions. For the diet intervention in obesity, you separated responders and non-responders, but actually, it looked like both groups lost significant amounts of weight. In obesity, it is often the long-term maintenance of weight loss that is problematic. I don't think it is good to lose weight in the short-term if this is not maintained. It may be better to lose less weight and then maintain it.

Tesson: This study was done for 32 weeks. For the first six weeks, the patient had a strict regimen, and then there was a period of weight stabilization for 26 weeks. The mean weight loss after 6 weeks was about 50% greater in the diet responsive group.

Martinez: Are you sure these are not cheaters and non-cheaters?

Tesson: We started with more than 1100 individuals, and at the end we had only 353 patients who were compliant.

Martinez: The genetics involved could be the genetics of compliance.

Tesson: Compliance is very important in terms of phenotype. It is a major reason for poor response to diet; therefore every effort was made to take out noncompliant patients.

Martinez: The problem is then how applicable this will be to the population as a whole. Anything found in 15% who comply needs to be thought through carefully before it is applied to the other 85% that are unable to comply, even in the best possible circumstances which is a clinical trial. This is a problem we always have in clinical trials: how generalizable are the results of what we find in the population as a whole?

Rutter: This comes back to Nick's suggestion that we need to think of disease prevention. I wonder whether this is the right way to think about it. Nick, in his paper, was noting that risk factors are mostly dimensional, but more than that many other risk mechanisms apply in people who do not yet have the disease. We need to think about the implications in public health terms across the whole population and not just in terms of disease groups.

Tesson: I agree, but the first goal is to find an effect. Usually, the effect of the gene polymorphisms, even if you look at a population of diseased individuals, is pretty small. If you want to see something you might have to address your questions to a population that might give you answers, which is likely to be the disease group. Then you can always translate that into a larger population.

Rutter: Almost in a throwaway line in your conclusions, you introduced the need for considering epigenetic mechanisms. Can you say more about this?

Tesson: Epigenetics is really important, and to my opinion, should now be considered in disease etiology. It's a rapidly evolving field. For example, in the case of hypertension in rats there is recent evidence of a direct link between the maternal diet, AT1 gene methylation status and the development of the disease later on in life (Bogdarina et al 2007). The epigenetic modifications are supposed to be reversible, and the mutations are not. This is important. There are many possibilities for looking at the rate of methylation of the genes found to be associated with the disease. In my lab, we are looking forward to starting the study of the role of epigenetics in G × E.

Robertson: PPARγ has been implicated as a partaker in gene × gene (G × G) interactions with regards to susceptibility to diabetes (Savage et al 2002). This is one of the few examples of this in human genetics. Are you configured in any way to re-examine the strength of this relationship or its attributable effects? There are so few demonstrable examples of a G × G interaction in human disease states.

Tesson: Do you mean in the study with the PPARγ and ACSL5? We looked at G × E and didn't find any. We had a trend, but it was not significant. ACSL5 is supposed to be a target gene for PPARγ so we thought we stood a good chance of finding a synergy. To look at this we would need large populations; on a larger population we might find significance.

Snieder: What will your main outcome phenotype be in your salt-sensitive hypertension study? Will you look at the actual blood pressure levels?

Tesson: Yes, 24 hours AMBP will be performed and blood pressure will be considered as a continuous variable for calculations.

Snieder: The Chinese study (Gu et al 2007) is also funded to do a genome-wide association study. Your study will have to compete with that one.

Tesson: These are the rules of the game!

References

Bogdarina I, Welham S, King P, Burns S, Clark A 2007 Epigenetic modification of the renin-angiotensin system in the fetal programming of hypertension. Circ Res 100:520

Gu D, Rice T, Wang S et al 2007 Heritability of blood pressure responses to dietary sodium and potassium intake in a Chinese population. Hypertension 50:116–122

Savage DB, Agostini M, Barroso I et al 2002 Digenic inheritance of severe insulin resistance in a human pedigree. Nat Genet 31:379–384

GENERAL DISCUSSION II

Poulton: I have a question for people doing genome-wide association studies. There are problems with false positives using this approach. How does one begin to grapple with this considerable potential for false positives?

Martin: Replication is the only answer.

Uher: Out of the replications for type 2 diabetes, three out of nine genes were replicated, and at a much more lenient level of statistical significance (Zeggini et al 2007).

Martin: You always expect replications to have a lower level of significance, because you have capitalized on chance in the initial findings. It's regression to the mean.

Uher: In the replication study, if they start off with nine single nucleotide polymorphisms (SNPs) coming from a previous study, shouldn't the significance value be adjusted at least for these nine multiple tests?

Rutter: One of the things that was different about the Wellcome Trust study was that they built in replication within their sample from the very beginning. They also had a number of other checks to try to look at the validity of their findings.

Martin: It is becoming clear that you'll need huge numbers for this. Having promoted the idea of using monozygotic (MZ) twins, it is hard enough to find replicable mean effects. To find replicable variance effects the power is much lower again because it is a variance ratio test rather than a *t*-test. It is worth doing, though. We already have 2000 MZ pairs, and we need to scrape up another 10 000 around the world. Given all the twin registries we can probably do this. Hearing Frederique's talk, I was thinking how wonderful it would be if the sodium interventions were in MZ twins! This would be a much tighter experiment, to see which pairs were most variable in their response. Another point that has been touched on today is that these effects are very small. But this is across the whole population. At the moment we have no way of partitioning the population. My belief is that in the fullness of time we will find that many of these genes have much larger effects. At the moment we have to look at the average effect diluted out by all the families in which there is no effect, perhaps because of a G × E or G × G interaction.

Rutter: It also comes back to the need to think about risk in a variety of ways and to express in terms that are understandable (see Academy of Medical Sciences 2007). For example, the relative risk of a baby having Down syndrome in women over the age of 40 is 16 times that in women aged 20–25. This sounds a huge risk. However, the absolute risk is only 1%. If we look at total population attributable risk, the correlation between Down syndrome and IQ is 0.076. This sounds so trivial as not really being worth bothering with. But in individuals with Down syndrome this reflects a deficit of 60 IQ points. All of these are true, but too often people talk about risk as if there was a single figure that meant things. One needs to think about risk that may be low in population terms, but very high in those who are actually affected by whatever it is you are looking at. Interestingly, there is now a chair at Cambridge on the public understanding of risk.

Kleeberger: One of the points I wanted to raise is that while we are talking about gene–environment interactions, we have only spent a little time on the environment, and the assessment of environmental exposures. Getting accurate estimates of exposures is difficult, whether this is air pollution or toxins in our food and drink, but these are important questions.

Rutter: That is an important point. From the twin study data it is clear that environmental effects account for quite a lot of the variance on all the multifactorial disorders. Yet the kinds of measures that are used aren't terribly solid. They include broad thing such as socio-economic status (SES). Even where there are good measures the care taken in testing for environmental mediation is usually poor. If we are to study gene–environment interactions we have to be studying the environment properly, as well as the genes.

Martin: I'd like to challenge that. It depends from case to case, but in research on schizophrenia, if we look at the Fuller Torrey study (Fuller Torey et al 1994) in which they recruited MZ twins discordant for schizophrenia, they measured everything possible and there wasn't a single factor that reliably predicted risk. It is the 'God of gaps' issue: we could always claim that they didn't measure the right things, but it is not that they didn't try. Look at IQ and scholastic ability back in the 1960s, where Loehlin & Nichols (1976) did a careful analysis of 850 pairs of twins who had sat the US National Merit Scholarship Qualifying Test. They asked 960 questions about the environment trying to assess what could account for MZ discordance. Once again, they found nothing at all. Robert Plomin conducted a large study (Chipuer & Plomin 1992) in which he measured the home environment carefully. The only variables he found important turned out to be largely genetic in origin, anyway. I don't think it is fair to say that people haven't worked hard at the environment. Nor do I agree with Steven Kleeberger when he says that no one here is interested in measuring environmental risk factors. Most of us are doing our best.

Kleeberger: I did not mean to imply that no one is interested in measuring environmental risk factors; it is just that there is no one here who specializes in that.

Rutter: Of course, things have been done, but they are mostly Mickey Mouse measures, they are not theory driven, and for the most part they are not biologically plausible. Moreover, they have stuck with the postnatal environment, with little attention to the prenatal environment. Very little attention has been paid to developmental perturbations. For example, there is the recent Reichenberg et al (2006) study looking at paternal age and the risk for autism. It showed a huge effect and two other studies found the same. Is that an environmental effect? In a sense it is, but it is not a measured environment on the individual. It is much more like the risk for Down syndrome in relation to maternal age. I would still argue that we haven't done this as thoroughly as we should have done.

Martin: MZ twin discordance is the big puzzle for all these complex traits. Not a single one has an MZ concordance of over 50%. The obvious thing is to look for the measured environmental discordances, and schizophrenia was a case where people tried hard and failed. This is what has driven my interest in epigenetics. We and others are now starting to do methylation chips on discordant MZ pairs.

Dodge: There is a big difference between measuring the environment a lot and measuring it well. We measure the environment a lot, but most of the measures are inherently flawed because they are self-reports about the environment, rather than direct measures. We have few studies that directly measure the environment. I would argue that we are going to introduce all kinds of biases in this way.

Martinez: I'd like to add that I am involved in a longitudinal study that is now 25 years old, from birth. The further away you get from the exposure the more the report is biased with respect to the assumed consequences of the exposure. Individuals who have the assumed consequence of the exposure often bias their reports with respect to whatever they think the exposure caused to them.

Koth: A couple of years ago NIH assembled a group of experts to look at genetic association studies. The issues of n and the environment came up, of course. One of the things this panel recommended was developing standards for defining and logging clear and specific clinical criteria and outcomes, developing standard operating procedures for how clinical information and patient material are to be assessed, and developing standards for including relevant environmental factors and how these would be measured etc. Without standardization and cleaning of the clinical sample it can be very hard to compare results from different studies.

Snieder: There is the Public Population Project in Genomics or P3G (*www.P3G-consortium.org*). This is a loosely organized association of all the population projects in the whole world. One of the major issues is trying to harmonize all the phenotypes being collected in these population studies. They only include population studies larger than 10 000 individuals.

Koth: All of this has to be entered into a database that will be used to analyze the phenotype to genotype analysis in a standard way.

Snieder: I agree: there needs to be harmonization of what is collected. The way that DNA is being collected is also harmonized.

Rutter: It is not just standardization we need; we also need agreement on how we are going to analyze the data. One of the fights in the autism consortium was that many people wanted to be able to analyze their own way, in whatever fashion they chose. In the end, this got stamped on, on the grounds that the chances of false positives are enormous if everyone is doing different analyses. The agreement was that there had to be a data analysis plan that was agreed in advance in order to avoid data dredging. Of course, this is not saying that if some unexpected new finding comes up you don't change your plan, but it is saying that you have to specify in advance what you are going to do. If you change from that, you must specify why and how you have changed the analytic plan. This hasn't happened to anything like the same extent in environmental studies, but it applies just as strongly. There is a need to take this forward. One of the down sides of this, which from a science point of view seems to be incontrovertibly correct, is what it does to careers. If you have publication by consortia (which is necessary), there is the problem of people's careers needing to be based on having first author publications, how do we deal with this?

Martin: This is exactly the question I asked at a recent NIH meeting: what happens to the junior postdocs who are the ones who actually do the work? No one had a plan as to how to look after them.

Reeve: Biology is at the stage nuclear physics was 50 years ago where they had huge teams. I don't know how they got round the problem, but we do have a problem of career development in this new age.

Rutter: The young investigators are our investment in the future. If they don't have the chance of taking things forward with their own ideas we are in trouble.

Braithwaite: In physics and mathematics, all the authors are listed alphabetically.

Rutter: That's why so many people change their names from Zachariah to Abrahams!

Dodge: If there is a public policy from this work that makes sense to me, it is to develop universal measures of the environment that are by matter of practice tracked in individuals. Particularly with regard to prenatal care and early life environments, we need the candidate environmental variables that some group deems to be potentially important. It could be a policy practice to record measures of important environmental variables in archived birth records and documents. It would be one step towards better measurement of the environment.

Rutter: Let me ask another question. We touched on the sensitivities of ethnic minority groups to genetic research. It is understandable that this exists because of the history of eugenics. But on the other hand the importance of using variations among ethnic groups as a way to understand causal mechanisms has to be taken fully on board. The obvious example from psychiatry is the increased rate of schizophrenia and other psychoses in Afro-Caribbeans living in the UK or the Netherlands but not in those living in the Caribbean (see Rutter & Tienda 2005). This is a robust finding and it has to imply some kind of environmental factor. It may well involve G × E interactions, but it is not simply a migration of the genetically at risk. Similarly, there are curious variations among ethnic groups in both the USA and the UK, where risk factors that appear influential in one ethnic group don't apply in others. Thus, the apparent effects of corporal punishment in Caucasians do not seem to apply to African Americans. Similarly, with respect to educational attainment in both the US and USA, single parenthood is a risk factor among Caucasians but not among African-Americans or Afro-Caribbeans in the UK. If we understood why there were these ethnic variations, it could carry messages for mechanisms that could apply more broadly. From a policy point of view we need a public engagement exercise to convince people that we are not in the business of labeling groups as genetically inferior or superior. We are in the business of trying to understand mediating mechanisms for risk in relation to all sorts of outcomes, good, bad and indifferent.

Koth: I think the more genetics proves it can help the development of new therapeutics or cut down on adverse effects to drugs or vaccines, the more people will have a better attitude towards genetic studies.

Rutter: The fact that there are ethnic variations in response to certain forms of treatment is reasonably well established. But it seems to me that the message should not be the need for different treatments for different groups. That would not work because genetic mixture is so great in most of these ethnic groups. You need to go the DNA in order to sort out whether they are responsive of not responsive to whatever drug one is looking at. This is where there needs to be a clear understanding that ethnicity is a complicated concept: it is in part a biological concept; in part what society does in labeling people; and in part self-identity. In so far as it is biological, genetic mixture is high in almost all populations in Europe and the USA.

Koth: If the image of genetics is that it is used to protect these groups, then there would be more acceptance.

Rutter: Yes, and it is unfortunate that we have esteemed colleagues who put things in simplistic and misleading ways (see Bhattacharjee 2007).

Dodge: This issue is compounded in the USA by the unfortunate fact that African-Americans are arrested at a higher rate than other groups, and DNA is now used to prove guilt in crimes at a higher rate. Put these emerging trends together with the eugenics of the past, and we have two factors that conspire to make many people from minority ethnic groups highly skeptical about the worth of collecting DNA and DNA research. There has to be a concerted effort to solve this problem.

Martinez: Let me suggest that this will be solved when these communities acquire control of the studies being done. Once the situation is that a significant number of native American scientists, for example, start becoming involved in the studies, then there will be greater acceptance.

References

Academy of Medical Sciences 2007 Identifying the environmental causes of disease: how should we decide what to believe and when to take action? Academy of Medical Sciences, London

Bhattacharjee Y 2007 Public policy: Watson condemned for comments on intelligence. Science 318:550

Chipuer HM, Plomin R 1992 Using siblings to identify shared and non-shared HOME items. Br J Dev Psychol 10:165–178

Fuller Torrey E, Gottesman II, Taylor EH, Bowler AE 1994 Schizophrenia and manic-depressive disorder: the biological roots of mental illness as revealed by a landmark study of identical twins. Perseus Books Group

Loehlin JC, Nichols RC 1976 Genetics and personality: a study of 850 sets of twins. University of Texas Press, Austin, xii, 202 pp

Reichenberg A, Gross R, Weiser M et al 2006 Advancing paternal age and autism. Arch Gen Psychiatry 63:1026–1032

Rutter M, Tienda M 2005 Ethnicity and causal mechanisms. Cambridge University Press, Cambridge

Zeggini E, Weedon MN, Lindgren CM et al 2007 Replication of genome-wide association signals in UK samples reveals risk loci for type 2 diabetes. Science 316:1336–1341

8. Longitudinal studies of gene–environment interaction in common diseases—good value for money?

Stephen P. Robertson and Richie Poulton*

*Clinical Genetics Group, Department of Paediatrics and Child Health, Dunedin School of Medicine, University of Otago, Dunedin, New Zealand and * Dunedin Multidisciplinary Health & Development Research Unit, Department of Preventive & Social Medicine, Dunedin School of Medicine, University of Otago, Dunedin, New Zealan*

Abstract. Prospective cohort studies are costly and time consuming yet appear to be the best means for understanding how genes interact with environmental risk factors to cause disease. This information is a necessary prerequisite for evidenced-based disease prevention, yet not all researchers agree about the importance of studying the interplay between genes and environments. They argue that we already know enough about which environmental 'exposures' can prevent most common diseases, for example, wholesome diet, adequate housing/income and access to good healthcare. Implicit is the notion that current disease categories (i.e. phenotypes) are 'real' and represent homogenous entities, and that identifying environmentally mediated risk is relatively straightforward. Other concerns relate to scientific basis, utility and ethics. These arguments are critically examined for a range of disorders, from diabetes, cancer and inflammatory bowel disease to depression. We refute the contention that incorporating the measurement of genotype into longitudinal-epidemiological studies is wasteful or unlikely to yield significant benefits.

2008 Genetic effects on environmental vulnerability to disease. Wiley, Chichester (Novartis Foundation Symposium) p 128–142

Slow progress understanding the genetic basis of many common diseases has been attributed to methodological factors and/or misconceptions surrounding the genetic architecture of disease susceptibility (Davey Smith et al 2005). Even the latest, largest and most comprehensive case-control study has only managed to identify genetic components of disease susceptibility that explain a small fraction of population attributable risk, and only for some of the conditions investigated (The Wellcome Trust Case Control Consortium 2007). These disappointing results are based on the standard genotype-phenotype association (i.e. main effect) approach, and underscore the potential importance of studying gene-gene and gene-environment interactions. Ideally this should be done using longitudinal designs that are capable of recording, with some precision, environmental exposures across the lifecourse (e.g. Collins 2004, SACGHS 2007).

However, not everyone is convinced that this is the best way forward. For example, Chaufan (2007) has questioned the value of further investment in large, prospective-longitudinal cohorts, and raised concerns about the scientific basis, utility and ethics of such studies. Her main objection is that we already know enough about which environmental factors underlie the most common diseases, and that many of these are already modifiable Implicit in this argument is that new medical breakthroughs are unnecessary for achieving significant reductions in disease burden and that marrying information about genetic variation with known environmental risks will be of little value to interventionists and policy makers (Cooper & Psaty 2003).

We disagree with this position and in what follows describe how information about genes *in the context of their environment* can improve: (i) our understanding of basic pathophysiology; (ii) definition of disease phenotypes; and (iii) uncover new environmental contributors to disease states. In so doing we reiterate much of the previously developed justifications for longitudinal gene–environment (G × E) studies (Khoury et al 2005) but also re-examine these ideas in the light of recent empirical findings.

Type 2 diabetes mellitus as an illustrative example

The persuasiveness of Chaufan's argument comes from her dependence on type 2 diabetes as her main illustrative example. It is true that environmental factors can account for up to 80–90% of the population attributable risk for this condition (Cooper & Psaty 2003), and it may be that in a profoundly diabetogenic environment such as exists in many 21st century developed countries, knowing about G × E interactions adds little *per se* to the management of an overweight and inactive population. If an environmental contributor is near ubiquitous and the genetic predisposition common as well, interventions are most sensibly weighted towards environmental risk factor modification.

Even here, though, there is room for further research, since the etiopathogenesis of type 2 diabetes may not be as well understood as some suggest. Specifically, Chaufan implies that dietary intervention to prevent prenatal 'programming' leading to susceptibility to develop type 2 diabetes (the fetal origins of adult onset disease hypothesis) is as evidence-based as dietary management of the adult diabetic state. However, many questions remain in this area. At what time in gestation and what nutritional/hormonal factors are involved in this programming? Is such susceptibility determined genetically? If so, is it the mother or the fetus that bears that predisposition?

Some of the emerging answers are surprising and counterintuitive, indicating for example that periconceptual nutritional factors may play a role, as distinct from mid-gestational endocrine and metabolic influences (Gluckman & Hanson 2004). Later on in gestation, evidence suggests that fetal and maternal genetic factors contribute to birth weight (Hattersley et al 1998, Frayling & Hattersley 2001). It is plausible that such factors may also operate at the very beginning of the human lifecourse but their identity, and the environmental factors they synergize with, remain unknown (Bloomfield et al 2006), awaiting discovery.

Chaufan also makes a strong case that inequalities in the provision of health care and education are compounding the growing problem of type 2 diabetes in the developed (and increasingly, less developed) nations today (Chaufan 2007). This is an important point, and one with which we agree, but it is concerned primarily with issues about resource allocation and distributive justice. This does not negate the value of studying how the combined effects of genes and environments predispose to ill health.

Improving our understanding of basic pathophysiology

Among the variables examined and controlled for in epidemiological studies studying risk factors for various disease states, many are unmodifable—sex, stature, skin pigmentation, birth order and ethnicity all being examples. Their unmodifiable character has not prevented their inclusion as risk factors and co-variables for the simple reason that controlling for them can bring other risk factors to attention during analysis. Evidence for a main effect (the association of a variable or risk factor to the disease state at a statistically significant level over the entire study population) is frequently absent initially; the variable being sought only comes to attention during stratification analysis as being significant for a fraction of the study sample. Why should stratifying for sex, for example, be a legitimate and powerful tool (Rutter et al 2003), but accounting for the effect of measured genotype be considered a folly? Genetic factors should be treated no differently to sex, ethnicity, birth order and the like. Commonly arising genetic variants in the population can define significant sub-populations that an environmental risk factor can operate on to predispose to disease with that same risk factor not conferring risk to those of a different genotype.

There is ample precedent for the value of this general approach, as several examples from the Dunedin Multidisciplinary Health and Development Study illustrate. In 2002, we showed how adult violence and antisocial behavior was predicted by an interaction between childhood maltreatment and a functional polymorphism in the promoter region of the gene encoding the enzyme monoamine oxidase A ($MAOA$) (Caspi et al 2002). This was the first report to demonstrate the operation of G × E for a behavioral disorder using a measured gene, and did so using comprehensive information collected throughout the first three decades of participants' lives. An important contribution of this study was to demonstrate that G × E do indeed operate in psychiatry, much as they do in other branches of medicine. This finding challenged the conventional wisdom of the day, which was that G × E were probably rare and of little importance in the behavioral sciences.

In a second study using the same basic design and approach, we showed how a functional polymorphism in the promoter region of the serotonin transporter gene (*5HTTLPR*) interacted with life stress to predict depression (Caspi et al 2003). Specifically, the *5HTTLPR* short allele conferred risk in the presence of life stress whereas carriers of the long allele appeared resilient to adverse life events. After ruling out gene–environment correlation we took advantage of the unique attributes of the longitudinal design to address the possibility of spurious findings due to gene–'gene' interaction via postdiction analysis. This involved testing whether depression that occurred at least three years prior to life stress (measured between ages 21 and 26 years), also interacted with *5HTTLPR*. It did not, thereby supporting the original G × E hypothesis. Further, with the longitudinal database we were able to test the idea that some genes work to modulate responsiveness to the environment. We reasoned that if this were the case, then the G × E should be present regardless of the timing of the stress exposures. In support, we found that both proximal (i.e. adult) and distal stress (from the first decade of life) interacted with *5HTTLPR* to predict depression in early adulthood.

Both these studies highlight a key benefit of collecting genetic information in developmental-epidemiological studies. This is captured nicely by Moffitt and colleagues (Moffitt et al 2006) who state that, 'The very special gift from a reliable G × E finding is clear evidence that there must be a pathway of causal processes connecting the three disparate "end points" forming the gene-pathogen-disorder triad.' (p.18.) The key point here is that these linkages may not be obvious, but knowing that three end-points are connected increases the likelihood of finding paths (or mechanisms) that unite them. This is of particular value in circumstances where environmental risks exhibit delayed effects, especially when risk exposure precedes the emergence of disorder by many years, as is often the case for mental disorders. Knowing that this depends on a particular genetic locus can help guide mechanistic explorations in the laboratory.

A third such study showed how the development of psychosis following exposure to cannabis during adolescence was moderated by a functional polymorphism in the catechol-*O*-methyltransferase (*COMT*) gene (Caspi et al 2005; see Fig. 1). Interestingly, this latter association was age-dependent—cannabis use in young adulthood did not elevate risk for developing psychosis, even in the pre-sence of genetic vulnerability in the form of the *COMT* valine allele—again emphasizing the value of longitudinal designs for elucidating potential

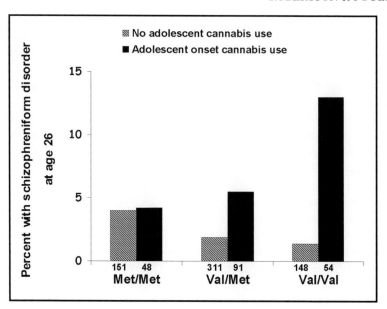

FIG. 1.   The association between adolescent-onset cannabis use and adult schizophreniform disorder is moderated by variation in the *COMT* gene. Adolescent cannabis use was defined as those who had used cannabis in the past year at age 15 (15% of the sample) or those at age 18 who had used cannabis at least once per month in the previous year (17%). Of the total sample, 26% were classified as adolescent-onset cannabis users. For details about schizophreniform diagnosis at age 26, see Poulton et al (2000).

mechanisms. Additionally, determining the timing and magnitude of the risk posed by cannabis for psychosis can help inform a sometimes heated debate about the harms associated with cannabis use. A sound evidence-base is a necessary pre-requisite to policy that balances citizens' rights and autonomy while protecting their health. Because cannabis sits at the nexus of health and legal policy it presents special challenges to lawmakers and public health officials. Thus knowledge about the *COMT* gene can help us to understand the basic pathophysiology of schizophrenia, thereby opening up new treatment possibilities, as well as informing national drug and health policy and legislative reform (Fergusson et al 2006).

Similar examples can be found for physical diseases. For example, Inflammatory Bowel Diseases (IBD), comprising a group of pathologically heterogeneous conditions with significant genetic underpinnings, have long been considered to be disorders of immune dysfunction. To date, treatment approaches have been generic and non-specific, largely owing to a lack of precise understanding about the nature of the immune defects. The recent identification of several genetic variants in genes encoding components of the innate immunity pathway conferring susceptibility to these conditions has significantly improved understanding of the pathophysiology of the condition. These findings have refined understanding the nature of the dysregulated relationship between the gut immune system and intestinal microflora and placed G × E as central to the pathogenesis of IBD. Further understanding will require measurement of the environment—for example the impact of composition of the gut microflora and host nutritional factors, including diet—and the way that they impact on the activity of the disease (Ferguson et al 2007). It would be nonsensical to study these factors independent of the various predisposing genotypes that have shown main effects in genome screens to date.

Genes can improve phenotyping of complex disorders

Among multifactorial psychiatric disorders Attention Deficit Hyperactivity Disorder (ADHD) stands out due to its marked heterogeneity in clinical presentation. Children with a diagnosis of ADHD vary in a number of important ways including: intellectual function; presence/absence of co-morbid conduct disorder; differential response to stimulant drug regimens; natural history (course) of symptoms; and in long-term prognosis. This begs the question: do the different features (and combinations thereof) have different etiologies? And if so, is the current ADHD diagnosis mixing apples with oranges, and conflating several distinct disorders that would benefit from quite different treatment approaches? This question is particularly relevant in disciplines like psychiatry for which diagnosis, in the absence of physical tests, relies upon symptom syndromes. Recent research has shown that polymorphisms in dopamine genes (*DRD4* and *DAT1*) are associated with differences in intellectual functioning among children diagnosed with ADHD, after controlling for the severity of symptoms. These same genes also predicted who was at risk for the worst outcomes in adulthood (Mill et al 2006). A more recent study in three different samples provided evidence for molecular-genetic subtyping of antisocial behavior within ADHD (Caspi et al 2008), reinforcing how genes can act as effect modifiers (Fanous & Kendler 2005). Together, these studies demonstrate how genetic information can be used to resolve clinical heterogeneity and refine diagnosis (Krishnan 2005).

Proof in principle of how knowledge about genetic variability can inform psychiatric nosology, and might even have clinical testing utility, as well as elucidating pathogenesis, emphasizes the value of genetic information. There are parallels in respiratory medicine with respect to the asthma phenotype, another clinical syndrome marked by considerable heterogeneity (Wenzel 2006). Genes can also help improve phenotypic definition in other branches of medicine. For example, the cancer field is beginning to witness the emergence of a new (yet to be fully realized) disease taxonomy based on molecular profiles, with greater precision and subtyping of diseases that foreshadows more effective treatment (Potter 2005).

When is an environmental risk factor really an environmental risk factor?

Gene–environment correlation (rGE) describes how individual genetic differences can 'drive' differential environmental risk exposure. In other words, exposure to environmental risk is not a random phenomenon; rather it stems in part from differences in genetic endowment (Plomin et al 1977). rGEs come in three main forms: (i) Passive rGE refers to environmental influences linked to genetic effects that are external to the person. For example, parents create the early child-rearing environment, as well as providing genetic material to their offspring; (2) Active rGE which, in contrast, can take the form of selecting specific environments or 'niche picking'; and (3) Evocative rGE which arise largely from factors within the person (Rutter et al 2006). The key observation for the current discussion is that environmentally mediated risk cannot always be assumed, despite appearances. From a public health perspective, the independence of environment from gene may not seem especially important—whatever the genesis of the environmental risk, the effects are presumably the same. However, this overlooks the possibility that unmeasured genes 'driving' the environment, via intermediate behaviors, may be suitable targets for intervention. New, albeit more distal targets are of value for no lesser reason than the effectiveness of current interventions for many chronic diseases is moderate at best.

Confirming the nature of environmental risk

Establishing biological plausibility brings confidence to public health initiatives that address risk factors that have modest main effects in conferring susceptibility to common disabling conditions. Hunter (2005) reviewed the example of the association of red meat with colorectal cancer and the subsequent observation that variation within the *NAT2* gene implicates heterocyclic polyamines as the responsible carcinogenic moiety. Such reductionistic insights have the potential to inform studies aimed at detecting other potential sources of environmental carcinogens— a formidable challenge if methodological approaches were restricted to conventional case-control study designs owing to the complexity of environmental exposures under study (Rothman et al 2001). Another example of how genetic information can help explain well known but poorly understood findings concerns the association between breastfeeding and IQ, in which moderation by the *FADS2* gene implicates fatty acids in breast milk as the likely mechanism explaining this association, as opposed to the nurturing and bonding that accompanies the act of breastfeeding (Caspi et al 2007).

Wither genotyping?

It must be emphasized that the detection of G × E does not necessarily imply that the appropriate intervention will require the genotyping of the population in which that intervention will be applied. For example, the consumption of red meat is shown to exhibit a moderate main effect on the risk of developing colorectal cancer, and subsequent work has shown that the most strongly predisposed fraction of the population are defined by the possession of 'fast metabolizer alleles' in the gene *NAT2*. Under such circumstances, it would be reasonable to promote an intervention to the population as a whole (such as moderation of red meat intake) as a reasonable preventive strategy if no significant harm accrued to the 70% of the population in whom the strategy would be less effective. The alternative (not conducting G × E studies to identify such strategies), is to fail to identify (or validate) a modifiable factor that significantly affects mortality and morbidity. The inclusion of genetics in the epidemiological studies has enhanced the science underpinning the public health intervention.

The use of family history to identify who may be at risk of an adverse G × E outcome is increasingly being viewed as a cost effective and feasible tool with which to apply these new findings (Scheuner et al 1997, Khoury et al 2005). Moreover, the demonstration of G × E does not mean that genotype needs to be 'modified', or even measured—it is clear that it is the environment that needs to be addressed. Although gene therapy may represent a promising approach for rare and very disabling genetic diseases, few would argue that genetic modification on a large scale represents a plausible approach to control of common disease states in the 21st century. In a health system that is increasingly demanding cost efficiency, the utility of targeted management (whether it be pharmaceuticals, lifestyle interventions, or nutritional management) can only mean a more efficient use of the health dollar and sparing individuals needless or less fruitful interventions. To posit that all interventions can be homogeneously applied, with equal weight, to a population with a disorder that is causally heterogeneous, implies some virtue in therapeutic imprecision.

The real danger in being nihilist about the value of studying G × E using prospective-longitudinal methods is that modifiable environmental factors will be dismissed as irrelevant simply because the relevant G × E interaction was not looked for using a sufficiently robust design. Efforts to find genetic factors conferring susceptibility to disease to date have mostly looked for main effects (The Wellcome Trust Case Control Consortium 2007). Emerging genetic epidemiology suggests that G × E can identify new risk factors; some of which may be modifiable. The implications for preventive public health strategies are obvious. The best method to discover them is via longitudinal studies that include the measurement of relevant genes in the population under study.

Historically, epidemiology, clinical trials and studies on the efficacy of pharmacological and non-pharmacological therapies have treated study populations as genetically homogeneous. This age began to close with the sequencing of the human genome in 2001, and the increasing affordability of wide-scale measurement of genetic variation will see its end. Substantial benefit can accrue from such an approach. Genotypes can be the lenses through which new environmental risk factors can be found, priorities on interventions refined and substantial health benefits accrued to all strata of society.

Acknowledgements

The work reported in this chapter from the Dunedin Multidisciplinary Health and Development Study was supported by a series of Programme grants from the Health Research Council of New Zealand; grants MH45070, MH49414, and MH077874 from the US National Institute of Mental Health, and grant GO100527 from the Medical Research Council of the UK. We thank Dr Phil Silva, founder of the Dunedin Study, and the study members, their families, their friends for their continuing support. Special thanks go to Professors Terrie E. Moffitt and Avshalom Caspi.

References

Bloomfield FH, Oliver MH, Harding JE 2006 The late effects of fetal growth patterns. Arch Dis Child Fetal Neonatal Ed 91:F299–304

Caspi A, McClay J, Moffitt TE et al 2002 Role of genotype in the cycle of violence in maltreated children. Science 297:851–854

Caspi A, Sugden K, Moffitt TE et al 2003 Influence of life stress on depression: moderation by a polymorphism in the 5-HTT gene. Science 301:386–389

Caspi A, Moffitt TE, Cannon M et al 2005 Moderation of the effect of adolescent-onset cannabis use on adult psychosis by a functional polymorphism in the catechol-O-methyltransferase gene: longitudinal evidence of a gene X environment interaction. Biol Psychiatry 57:1117–1127

Caspi A, Williams B, Kim-Cohen J et al 2007 Moderation of breastfeeding effects on the IQ by genetic variation in fatty acid metabolism. Proc Natl Acad Sci USA 104:18860–18865

Caspi A, Langley K, Milne BJ et al 2008 A replicated molecular-genetic basis for subtyping antisocial behavior in ADHD. Arch Gen Psychiatry, in press

Chaufan C 2007 How much can a large population study on genes, environments, their interactions and common diseases contribute to the health of the American people? Soc Sci Med 65:1730–1741

Collins FS 2004 The case for a US prospective cohort study of genes and environment. Nature 429:475–477

Cooper RS, Psaty BM 2003 Genomics and medicine: distraction, incremental progress, or the dawn of a new age? Ann Intern Med 138:576–580

Davey Smith G, Ebrahim S, Lewis S, Hansell AL, Palmer LJ, Burton PR 2005 Genetic epidemiology and public health: hope, hype, and future prospects. Lancet 366:1484–1498

Fanous AH, Kendler KS 2005 Genetic heterogeneity, modifier genes, and quantitative phenotypes in psychiatric illness: searching for a framework. Mol Psychiatry 10:6–13

Ferguson LR, Shelling AN, Browning BL, Huebner C, Petermann I 2007 Genes, diet and inflammatory bowel disease. Mutat Res 622:70–83

Fergusson DM, Poulton R, Smith PF, Boden JM 2006 Cannabis and psychosis: a summary and synthesis of the evidence. BMJ 332:172–175

Frayling TM, Hattersley AT 2001 The role of genetic susceptibility in the association of low birth weight with type 2 diabetes. Br Med Bull 60:89–101

Gluckman PD, Hanson MA 2004 Living with the past: evolution, development, and patterns of disease. Science 305:1733–1736

Hattersley AT, Beards F, Ballantyne E, Appleton M, Harvey R, Ellard S 1998 Mutations in the glucokinase gene of the fetus result in reduced birth weight. Nat Genet 19:268–270

Hunter DJ 2005 Gene–environment interactions in human diseases. Nat Rev Genet 6:287–298

Khoury MJ, Davis R, Gwinn M, Lindegren ML, Yoon P 2005 Do we need genomic research for the prevention of common diseases with environmental causes? Am J Epidemiol 161:799–805

Krishnan KR 2005 Psychiatric disease in the genomic era: rational approach. Mol Psychiatry 10:978–984

Mill J, Caspi A, Williams BS et al 2006 Prediction of heterogeneity in intelligence and adult prognosis by genetic polymorphisms in the dopamine system among children with attention-deficit/hyperactivity disorder: evidence from 2 birth cohorts. Arch Gen Psychiatry 63:462–469

Moffitt TE, Caspi A, Rutter M 2006 Measured gene–environment interactions in psychopathology. Concepts, research strategies and implications for research, intervention and public understanding of genetics. Perspect Psychol Sci 1:5–27

Plomin R, DeFries JC, Loehlin JC 1977 Genotype–environment interaction and correlation in the analysis of human behavior. Psychol Bull 84:309–322

Potter JD 2005 Epidemiology informing clinical practice: from bills of mortality to population laboratories. Nat Clin Pract Oncol 2:625–634

Poulton R, Caspi A, Moffitt TE, Cannon M, Murray RM, Harrington HL 2000 Children's self-reported psychotic symptoms and adult schizophreniform disorder: a 15-year longitudinal study. Arch Gen Psychiatry 57:1053–1058

Rothman N, Wacholder S, Caporaso NE, Garcia-Closas M, Buetow K, Fraumeni JF Jr 2001 The use of common genetic polymorphisms to enhance the epidemiologic study of environmental carcinogens. Biochim Biophys Acta 1471:C1–10

Rutter M, Caspi A, Moffitt TE 2003 Using sex differences in psychopathology to study causal mechanisms: unifying issues and research strategies. J Child Psychol Psychiatry 44:1092–1115

Rutter M, Moffitt TE, Caspi A 2006 Gene–environment interplay and psychopathology: multiple varieties but real effects. J Child Psychol Psychiatry 47:226–261

Scheuner MT, Wang SJ, Raffel LJ, Larabell SK, Rotter JI 1997 Family history: a comprehensive genetic risk assessment method for the chronic conditions of adulthood. Am J Med Genet 71:315–324

SACGHS (Secretary's Advisory Committee on Genetics, Health and Society) 2007 Policy issues associated with undertaking a new large US population cohort study of genes, environment and diseases. US, March 2007: US Department of Health and Human Services

The Wellcome Trust Case Control Consortium 2007 Genome-wide association study of 14,000 cases of seven common diseases and 3,000 shared controls. Nature 447:661–678

Wenzel SE 2006 Asthma: defining of the persistent adult phenotypes. Lancet 368:804–813

DISCUSSION

Heath: Thinking about the value of the cohort study, I fully agree we should be taking full advantage of all the prospective cohort studies, but unless we are working on a huge scale, our cohort studies are going to be limited to following findings that are emerging from genome-wide association studies, and basic science studies. Let's focus on your issue of stratification. Let's stratify by an environmental risk factor, for example early childhood trauma and maltreatment which can have potent long-lasting effects on depression risk. We mentioned that we would need perhaps 2000 depressed cases in our cell with early childhood trauma. So now we are thinking of something of the size of the 100000 individuals in the NICHD prospective cohort study. But I don't want to wait 20 years for those individuals to accumulate enough risk, and I doubt whether they'll get the detailed assessment of childhood maltreatment that we would want to see because of the mandated reporting requirements for child abuse in the different states in the USA. So let's not lose sight of the possibility of retrospective studies stratified by environmental exposure where we can accumulate 2000 depressed cases with assorted trauma as one cell in our design. Your argument is that if this stratification isn't done things can be missed.

Poulton: I am not saying that this is the only way to do good work. I am saying it is an important part. I see the value of studies like ours as being able to confirm the signals that come from these other designs. The concern is that these studies are perceived to be so costly, both in terms of money and also time, that people are frightened off them. For some phenotypes, you may not have to use this particular design, but for the ones that you and I work with, the fallibility of people's memory is such that retrospective work will always have limitations. I have been on a couple of panels recently assessing the viability of large cohort studies. The arguments from the funders have been that it is too expensive and takes too long, but the argument counter to this is that you can get an enormous and sometimes unpredicted return from these studies over time. If you really want to nail something down, this is probably the best way to do it.

Martin: You have a wonderful opportunity with your study to assess the accuracy of long-term recall of these risk factors.

Poulton: We published a paper in 1994 entitled 'Remembrance of things past' (Henry et al 1994). This showed that for certain actuarial type data, one can recall well after long periods, but for things like depression and mental states, they are extremely difficult to recall with any accuracy.

Martin: What about risk factors such as maltreatment?

Poulton: The conventional wisdom is that if the event is extremely salient and traumatic, one is more likely to remember it, and the lesser forms of trauma and milder forms of stress are lost in the mists of time. And there is also confounding by mental state and mood at the time of recall.

Martin: What are the correlations?

Poulton: You are asking about specific traumas: we didn't ask about this in our earlier study, but it's something we could look at. We plan to look at the accuracy of peoples' recall of socio-economic status. We have these data in hand, but haven't analyzed them yet. Many of the studies looking at the relationship between health and the social gradient are based on adults being asked about what their parents did when they were children. We have found that the correlation between socio-economic status (SES) at birth and age 15 in our group was only 0.5, so single measures have problems. If you are asking about a single measure 20 or 30 years prior, there might be substantial error

Martin: An interesting question would be the extent to which the correlation is with the actual event as opposed to the perceived or remembered event.

Rutter: The main problem is not that people who are ill remember things that didn't occur, but rather the other way round: we get under-reporting by people who don't have problems (see Hardt & Rutter 2004). It is no longer part of their thinking as to what is relevant. This results in a bias that is problematic. But I agree with Richie Poulton: this bias is much more influential on some things than it is on others. If you are dealing with attitudes and mental states, retrospective recall is hopeless. On concrete events it is much better, but they have to be memorable at the time. The answer is not to throw out retrospective recall as useless, but to be aware of the biases that can be built in. We have to remember that prospective studies also use retrospective recall. It may be over shorter times, but there is no way you can get around this completely.

Poulton: Nick Martin, earlier we were talking about breast feeding data that could be obtained retrospectively from your mothers. If the question is 'did you or did you not breastfeed', you can probably get reasonable data. But if you ask, 'did you breast feed for three or six months?' then I'd have less confidence in the accuracy of the data.

Martin: I know you are about to start a new wave of your study. It would be great to insert recall of these risk factors that are so important. It would be a service to the rest of us so we can know what is worth including in retrospective studies and what is not.

Poulton: This is planned, particularly with a focus on the antecedents to the physical health problems. The first paper was more focused on understanding the problems related to behavioral outcomes and their antecedents. Now that we are moving into the mid-life period, we are increasingly focusing on physical health outcomes. This will be the focus of a small set of questions which should be very generative, as you say.

Martinez: We have a good example of this recall issue. Bronchiolitis, a disease that happens in the first year of life in many infants, is strongly associated with subsequent asthma. We ascertained it in the first years of life and have been following these people to age 25 now. For the people who had bronchiolitis and now have asthma, their parents recall much better that they had bronchiolitis than those who don't have asthma now. It is at least twice more. Extraordinarily, some of these latter parents don't recall that they took their child to the doctor in the first year of life. One more issue: have you thought that the cannabis exposure to psychosis association could also be due to a common cause? Could it be that consuming cannabis regularly and being psychotic are caused by a common factor? It is not absolute with your data that it is cause and effect.

Poulton: I agree. We attempted the best controls we could. We controlled for co-occurring other substance use. We were fortunate in our study to have a child psychiatrist administering the first version of the Diagnostic Interview Schedule in 1982 when the children were 11, so we have good systematic data on mental health problems at age 11, including information about psychotic symptoms. Thus, we had the ability to control for early emerging psychotic-type symptoms before cannabis use started. The association was robust, and we looked for some biologically plausible, mechanistic explanations. There are data from animal studies and human experimental studies pointing towards mechanisms. It seemed that this is something worth pursuing. Others have similar findings.

Kleeberger: I have a quick follow-up question on the breast feeding study. You find that breast feeding has an effect on IQ later. Have you looked at other factors such as gender and breast feeding? Is the effect equally effective in males and females?

Poulton: The finding was that people who were breastfed (we didn't look at dose–response or length of time) who had a C allele on the *FADS2* gene (which is 90% of the population), had a 6–7 point difference in IQ compared with those who were not breast fed. That is about half a standard deviation. The GG homozygote, 10% of the population, had no benefit, but no disadvantage either. How much does half a standard deviation matter? Probably not that much on an individual level, but it is an interesting effect size looked at population-wide. The value of this study was twofold. At a theoretical level it showed that nature via nurture is probably the way things happen. And it was also to say that *FADS2* is involved. So what is it about the fatty acids that is involved in brain development? Is it something in terms of the child's genotype and how they process the fatty acids, or vice versa?

Kleeberger: Is there enough statistical power in the study to stratify on gender?

Poulton: In our case probably, and in the E-Risk case almost certainly. In terms of the things we controlled for, we had to rule out whether this association was a function of SES, or higher IQ in mums who breastfeed. Through the E-Risk study, we were also able to do something quite elegant, which is to ask whether this was a function of the child's genotype or the mum's genotype. We also controlled for length of gestation. Every study has its limits and we did what we could.

Dodge: I have a question about the nature of the gene–environment interaction process. You showed two familiar results. Maltreatment interacts with MAOA to predict conduct disorder, and maltreatment interacts with 5HTT to predict depression. There is a third result you didn't mention which is that stressful life events interact with 5HTT to predict depression also. Have you put these findings together into a single empirical analysis? The findings might suggest that child maltreatment is a blunt risk factor with multi-finality, which might lead to conduct disorder or to depression, depending on the genetic profile. Likewise the genetic profiles might be linked to disorders but only in the context of certain kinds of life experiences, either maltreatment or stressful life events. By putting these variables together into a single empirical analysis you might have a very strong empirical model predicting bad outcomes. Maltreatment might be a highly predictive factor in psychopathology, but the form depends on the genetic profile. Have you tried this?

Poulton: No. What we have done thus far is to test for specificity of gene and outcome. In the 5HTT promoter study, we looked for moderation by MAOA, and found no evidence for this. We also found specificity of outcome for the depression finding. Our test was a tough one using GAD. Given the known high levels of comorbidity between these two, this seemed pretty convincing. We didn't find specificity for the psychosis outcome. The cannabis–COMT interaction also applied to depression, suggesting that the biology underlying this association might be shared by psychosis and depression. It could, of course, just be a chance finding—time will tell.

Dodge: You report that 85% of those individuals who had experienced maltreatment and have a polymorphisim in MAOA end up with conduct disorder. That is impressive. What happens to the other 15%? Before we conclude that they emerge free of any diagnosis, it is plausible that if they have a polymorphism in 5HTT they will end up with depression.

Poulton: You are hitting on one of the limitations of a boutique study like ours. To do these more complex three or four-way interactions, we would be struggling with our sample size.

Rutter: As you say, this hasn't been done, but one query I put to Avshalom Caspi was whether the G × E life events finding could be explained by the maltreatment. The answer is no. After excluding all cases experiencing severe maltreatment the association between the short allele of the 5HTT promoter and depression was just as strong (Caspi, personal communication confirmed 17.12.07).

Martinez: Do depressed people maltreat their children? Could there be inverse causation? The gene predisposes to maltreating children and to becoming depressed.

142   ROBERTSON & POULTON

Rutter: It doesn't work like that. This is where we need quantitative genetics. The study that was done looking at corporal punishment on one hand and maltreatment on the other, also looked at depression as a further feature (Jaffee et al 2004). Let's start with the physical punishment and maltreatment. They sound the same, and people tend to assume they are. But from the quantitative genetics findings it is striking that whereas the effect of maltreatment is almost entirely environmentally mediated, the effect of corporal punishment is almost entirely genetically mediated, presumably through an effect of the child on the parent. What this study also showed was that although the effects of maltreatment and physical punishment were quite different, the parents who regularly used physical punishment were more likely later to use maltreatment. What started as punishment tended to escalate into abuse. With regard to maternal depression, it carries a risk for antisocial behavior in the children (Kim-Cohen et al 2005), but parental personality disorder is a much stronger predictor.

Martinez: The concern here is the same I have for the cannabis findings. You could have a set of genes or developmental conditions that predispose to maltreating people and becoming depressed at the same time. Then you would find the kind of apparent interactions that have been observed.

Rutter: Absolutely. This is a crucial methodological issue.

References

Jaffee SR, Caspi A, Moffitt TE, Polo-Tomas M, Price TS, Taylor A 2004 The limits of child effects: evidence for genetically mediated child effects on corporal punishment but not on physical maltreatment. Dev Psychol 40:1047–1058
Kim-Cohen J, Moffitt TE, Taylor A, Pawlby SJ, Caspi A 2005 Maternal depression and children's antisocial behavior: nature and nurture effects. Arch Gen Psychiatry 62:173–181

9. Gene–environment interactions in breast cancer

Kee-Seng Chia

Centre for Molecular Epidemiology, National University of Singapore, Singapore

Abstract. Breast cancer is one of the most frequently diagnosed cancers in women. It accounts for 23% of all cancers, with an estimated 1.15 million new cases in 2002. The role of the environment, such as reproductive factors, has been well studied in many epidemiological studies. Breast cancers also tend to cluster in families and are more common in monozygotic twins. Some of this clustering occurs as part of specific familial breast cancer syndromes where disease results from single alleles conferring a high risk. However, such alleles are rare in the population, and highly penetrant variants of *BRCA1* and *BRCA2* account for less than 20% of the genetic risk of breast cancer. The more common sporadic form of breast cancer are probably due to a polygenic inheritance of breast cancer susceptibility upon which environmental factors act upon resulting in breast cancer occurrence. Recent high-throughput genome-wide association studies are identifying several such genes, each with small absolute risk but with significant population level implications. The study of gene–environment interactions has thus far been confined to candidate gene approaches. The lack of large prospective cohorts with thousands of incident breast cancer cases makes the study of gene–environment interactions using a nested case-control design extremely difficult.

2008 Genetic effects on environmental vulnerability to disease. Wiley, Chichester (Novartis Foundation Symposium) p 143–155

Among the various cancers, breast cancer has probably attracted the most scientific interest. A cursory search for articles on 'breast cancer' on PubMed in October 2007 reveals 101 499 articles compared to 52 524 for 'lung cancer' and 35 424 for 'prostate cancer'.

Breast cancer is one of the most frequently diagnosed cancers in women. It accounts for 23% of all cancers, with an estimated 1.15 million new cases in 2002 (Parkin et al 2005). Globally, there is a four- to fivefold difference in the age-standardized incidence rates. International rates show breast cancer incidence is low in Asia, moderate in South America and Eastern Europe, and high in North America and Western Europe. Although these international variations could be partially due to registration factors, a comparison between registries with high indices of quality data confirms this pattern (Parkin et al 2005).

143

The importance of reproductive factors was first revealed in the 18th century by an occupational health physician, Bernado Ramazzini, who observed that nuns have higher rates of breast cancer compared with married women. In the 20th century, thousands of research papers confirmed the association between a woman's reproductive life and her risk of developing breast cancer. This was further confirmed by the increasing incidence following the introduction of oral contraceptive pill and hormone replacement therapy. Breast cancers also tend to cluster in families and are more common in monozygotic twins. Some of this clustering occurs as part of specific familial breast cancer syndromes where disease results from single alleles conferring a high risk, e.g. *BRCA1* and *BRCA2*. The more common sporadic form of breast cancer are probably due to a polygenic inheritance of breast cancer susceptibility upon which environmental factors act upon resulting in breast cancer occurrence.

Genetic susceptibility

Family history of breast cancer in first-degree relatives confers a 1.5–3.0 times risk compared to those without a family history (Greene 1997). Twin studies suggest that most of this familial clustering is due to genetic factors rather than shared environment (Lichtenstein et al 2000). Large population series of breast cancer cases with detailed family history can be used to estimate the contribution of rare but highly penetrant genes such as *BRCA1* and *BRCA2*. Such studies show that these genes account for less than 20% of the total genetic risk (Easton 1999, Peto et al 1999, Anglian Breast Cancer Study Group 2000). The remaining 80% could be due to yet undiscovered rare genes of significant effect or the combined effects of weakly predisposing alleles in many genes. Linkage studies will be ideal in the search *BRCA*-like genes. In the case of multiple genes with small effects, genetic association studies will be more appropriate (Risch 2000). If BRCA-like genes account for most of breast cancers, epidemiological evidence will show excess number of breast cancers in small number of families. However, most of the excess clustering of familiar cases is found to be distributed over many families, each with few cases (Antoniou et al 2002). This is supportive of a polygenic mode of inheritance.

The search for genes under a polygenic model has taken on two strategies: a candidate gene or a genome-wide approach. Both approaches have its supporters and detractors with no conclusive recommendation on a preferred approach in sight. The candidate gene approach has also evolved from studying a small number of genetic variants usually in the form of SNPs (single nucleotide polymorphisms) (Feigelson et al 1997) to more comprehensive approach using hundreds of tagging SNPs as the technology for high-throughput genotyping becomes more reliable and affordable.

In the genome-wide approach no prior knowledge of position or function of risk alleles is needed. In the human genome, there are about 10 million common SNPs with minor allele frequency (m.a.f.) of more than 5% (Kruglyak & Nickerson 2001). During meiosis, recombination tends to occur at distinct 'hot-spots', resulting in neighboring polymorphisms being strongly correlated (in 'linkage disequilibrium', LD) with each other. As a result, all the common SNPs could therefore be evaluated using a few hundred thousand SNPs as tags for all the other variants (Hinds et al 2005). Data from such tagging SNPs results in high dimensional datasets (large number of highly related data that far exceeds the number of subjects). Hence there is a very high risk of type 1 statistical error which require a statistically significant result to be raised from the conventional $P < 0.05$ to beyond $P < 10^{-5}$. In addition, statistically significant risk alleles will need to be replicated in other populations. As a result, most genome-wide association (GWA) studies are designed in stages with replication populations built in.

As an example, a recent GWA study was designed as a two-stage study of over 4000 cases and controls followed by a confirmatory stage on another population of more than 21 000 cases and controls (Easton et al 2007). In the first stage, 390 cases and 364 controls were genotyped for a set of 266 732 SNPs. Stage 2 consists of 3990 cases and 3916 controls genotyped for 12 711 SNPs selected on the basis of stage 1 results. The top 30 SNPs from stage 2 were further tested in 21 860 cases and 22 578 controls. Five novel independent loci demonstrated strong and consistent association with breast cancer ($P < 10^{-7}$). Of these, four contain plausible causative genes: *FGFR2* (fibroblast growth factor receptor 2), *TNRC9* (gene of unknown function with a tri-nucleotide repeat motif), *MAP3K1* (mitogen-activated protein kinase kinase kinase 1) and *LSP1* (lymphocyte-specific protein 1). This study underscores the usefulness of GWA studies to identify common susceptibility loci and the critical importance of large sample size and more dense SNPs coverage.

Environmental risk factors

Migrants who settle in countries with higher breast cancer incidence adopt the rates of the host countries, usually within one to two generations (Ziegler et al 1993, Jain et al 2005). In addition, countries that have undergone very rapid modernization are showing corresponding increase in breast cancer incidence (Parkin et al 2005, Parkin & Fernandez 2006). Rapidly falling birth rates are also associated with increase in breast cancer rates in the corresponding cohorts of women (Chia et al 2005). These findings suggest that environmental and lifestyle factors play a central role in breast cancer incidence.

There are several well-confirmed 'environmental' risk factors for breast cancer. These would include:

1. Delayed child-bearing
2. Early menarche and delayed menopause
3. Postmenopausal hormone use
4. Ionizing radiation exposure
5. Height
6. Benign breast disease
7. Mammographically dense breast
8. High body mass index (post-menopausal breast cancer only)

Unfortunately, most of these well-confirmed risk factors are not easily modifiable. Other possible but weaker risk factors like lack of lactation, physical inactivity, alcohol consumption and current oral contraceptive use can possibly be translated into cancer prevention strategies. While intervention with some of these factors may reduce breast cancer risk, some of these factors like hormone use are beneficial for other conditions like osteoporosis.

Internationally, there are differences in the age-specific incidence curves. In Western populations, the incidence is rare among those below 40 years of age and the peak incidence is in the post-menopausal age group. In stark contrast, many Asian populations display a pre-menopausal peak. This has prompted the hypothesis of breast cancer being two separate diseases: predominantly pre-menopausal in Asians whilst being predominantly postmenopausal disease in Caucasians populations (Waard 1979, Bray et al 2004).

In comparative studies between Asian and Western populations, age-period-cohort analysis showed stronger birth cohort effects in the Asian populations, namely Singapore (Chia et al 2005) and Taiwan (Shen et al 2005), compared to the Western populations, Sweden and the Caucasian Americans respectively. When the birth cohort was taken into account, there was no difference in the relative risk of breast cancer by age groups among Japanese and Western populations (Moolgavkar et al 1979) or Swedish and Singaporean women (Chia et al 2005). The difference in the age-specific incidence is potentially due to a time lag in exposure to lifestyle factors and thus, suggesting that over time, the incidence of breast cancer in Asia will increase as lifestyle factors approach that of Western populations (Chia et al 2005, Shen et al 2005, Leung et al 2002). With successive birth cohorts, the pre-menopausal peak is expected to move into the post-menopausal age group. However, there may be differences in exposure or response to genetic and/or environmental factors that may modulate the incidence among Asian populations. In particular, among the ethnic groups (Chinese, Malay and Indian) in Singapore, there appear to be ethnic differences in the temporal trends of breast cancer in spite of fairly similar declines in fertility rates (Sim et al 2006).

Gene–environment interactions

Apart from familial breast cancers associated with mutations in *BRCA1* and *BRCA2*, sporadic forms of breast cancers are most likely due to a complex interplay of the combined genetic effect of multiple weakly predisposing alleles and the known major lifestyle and environmental risk factors. In epidemiology, such interactions result in risk measures (rate ratios) that are greater than the sum of the individual factors. For example, the deleterious effect of delayed childbirth may be markedly different in subjects with a certain combination of predisposing alleles compared to those who do not have the risk alleles.

The ideal design for the study of gene–environment interactions is within a prospective cohort study (Manolio et al 2006). Women without breast cancer are recruited and exposure data collected. This would typically be questionnaire data but in this biobank era, would include biological specimens which could be used in future when new exposure biomarkers become relevant. The women are followed-up with repeated collection of questionnaire data and biological specimens. Once the cohort generated sufficient incident cases of breast cancers, these are compared to controls from the same cohort. Such nested case-control study overcomes the problem of recall bias and of selecting controls which does not come from the same study base as the cases. In the case of traditional case-control studies, it is possible to study gene–environment interactions if the retrospective exposure data to be collected can be free of recall bias; for example, age of first childbirth can be retrieved from birth registries. However, this may not be possible for the other major risk factors like age of menarche and menopause.

Most nested case-control studies have taken the candidate gene approach where a handful of known risk alleles are genotyped. For example, in a nested case-control design within the Singapore Chinese Health Study, the angiotensin I-converting enzyme (ACE) gene polymorphism modifies the protective effect of green tea consumption on breast cancer (Yuan et al 2005). Green tea polyphenols has been shown to exhibit antiproliferative and antiangiogenic effects in breast cancer cell lines as well as inhibit the size and multiplicity of carcinogen-induced mammary tumors and significantly increase the survival rate of carcinogen-treated mammals (Kavanagh et al 2001). Meta-analyses of epidemiological studies shows a 22% decrease in risk among highest green tea drinkers compared to non-drinkers or those of lowest intake (Sun et al 2006). Experimental data also showed that green tea polyphenols could inhibit angiotensin II-induced reactive oxygen species production (Ying et al 2003). It has also been shown that women with low-activity genotype of the ACE gene had a reduced risk of breast cancer compared with those possessing high-activity ACE genotype (Koh et al 2003). In the nested case-control study, two polymorphisms (*A-240T* and *D/I*) in the ACE gene were identified; those with the TT and/or DD genotypes were considered to be high-activity ACE genotype. In the group with high-activity ACE genotype, weekly or more frequent green tea drinkers have a 70% reduction in breast cancer risk whereas there was no protective effect among those with low-activity. This is the first epidemiological evidence that green tea may reduce breast cancer risk via the angiotension II/ NADPH oxidase-induced ROS/VEGF pathway in breast cancer.

In a nested case-control (1588 cases and 2600 controls) within the European Prospective Investigation into Cancer and Nutrition (EPIC) cohort, a haplotype approach was taken to investigate the role of common genetic variants in the acetyl-CoA carboxylase A (*ACC-A*) gene (Sinilnikova et al 2007). ACC-A is a key fatty acid synthesis enzyme that has been shown to be highly expressed in human breast cancer and other tumor types and also to specifically interact with the protein coded by one of two major breast cancer susceptibility genes *BRCA1*. In one common haplotype, homozygous carriers have higher risk (OR 1.74; 95%CI, 1.03–2.94). Another haplotype block showed interaction with menopausal status ($P = 0.016$): it was protective in pre-menopausal women (OR = 0.65, 95% CI: 0.48–0.87) but increases the risk in post-menopausal women (OR = 1.36, 95% CI: 0.62–2.99).

The genome-wide approach has yet to be used to discover gene–environment interactions. The sample size needed for such 'genome-wide interactions' will be much larger than gene discovery. Probably, most cohorts have yet to have sufficient incident cases for such an approach. The problem of multidimensionality will be greatly multiplied by the number of additional exposure variables. Eventually solutions to the computational and statistical challenges may be found but the biological understanding of the interactions may be more difficult to unravel.

Conclusion

In this post-genomic era, the pressure appears to have been towards a general discovery rather than a hypothesis-driven approach. One could argue that we do not know enough of the complexity to be able to generate meaningful hypothesis. Discovering the possible gene–environment interactions should lead to better understanding of the underlying mechanism and ultimately lead to directing preventive interventions to those who are susceptible (Bell 1998, Beaudet 1999). Others are less optimistic, questioning if testing for common genetic variants will have sufficient predictive power to be of practical use to identify high risk subjects (Vineis et al 2001). Unfortunately, scientific advancement and the progress of the biomedical industry do not occur in linear nor logical manner. Discoveries and inventions will outstrip each other in an accelerated and spiraling fashion. The scientific, business and regulatory communities must continuously engage each other to avoid jeopardizing the health of the population through premature release of biomedical products.

References

Anglian Breast Cancer Study Group 2000 Prevalence and penetrance of BRCA1 and BRCA2 mutations in a population-based series of breast cancer cases. Br J Cancer 83:1301–1308

Antoniou AC, Pharoah PD, McMullan G et al 2002 A comprehensive model for familial breast cancer incorporating BRCA I, BRCA2 and other genes. Br J Cancer 86:76–83

Beaudet AL 1999 1998 ASHG presidential address. Making genomic medicine a reality. Am J Hum Genet 64:1–13

Bell J 1998 The new genetics in clinical practice. Br Med J 316:618–620

Bray F, McCarron P, Parkin DM 2004 The changing global patterns of female breast cancer incidence and mortality. Breast Cancer Res 6:229–239

Chia KS, Reilly M, Tan CS et al 2005 Profound changes in breast cancer incidence may reflect changes into a Westernized lifestyle: a comparative population-based study in Singapore and Sweden. Int J Cancer 113:302–306

Easton DF 1999 How many more breast cancer predisposition genes are there? Breast Cancer Res 1:14–17

Easton DF, Pooley KA, Dunning AM et al 2007 Genome-wide association study identifies novel breast cancer susceptibility loci. Nature 447:1087–1095

Feigelson HS, Coetzee GA, Kolonel LN, Ross RIC, Henderson BE 1997 A polymorphism in the CYP17 gene increases the risk of breast cancer. Cancer Res 57:1063–1065

Greene MH 1997 Genetics of breast cancer. Mayo Clin Proc 72:54–65

Hinds DA, Stuve LL, Nilsen GB et al 2005 Whole-genome patterns of common DNA variation in three human populations. Science 307:1072–1079

Jain RV, Mills PK, Parikh-Patel A 2005 Cancer incidence in the south Asian population of California, 1988–2000. J Carcinog 4:21

Kavanagh KT, Hafer LJ, Kim DW et al 2001 Green tea extracts decrease carcinogen-induced mammary tumor burden in rats and rate of breast cancer cell proliferation in culture. J Cell Biochem 82:387–398

Koh WP, Yuan JM, Sun CL et al 2003 Angiotensin I-converting enzyme (ACE) gene polymorphism and breast cancer risk among Chinese women in Singapore. Cancer Res 63:573–578

Kruglyak L, Nickerson DA 2001 Variation is the spice of life. Nat Genet 27:234–236

Leung GM, Thach TQ, Lam TH et al 2002 Trends in breast cancer incidence in Hong Kong between 1973 and 1999: an age-period-cohort analysis. Br J Cancer 87:982–988

Lichtenstein P, Holm NV, Verkasalo PK et al 2000 Environmental and heritable factors in the causation of cancer—analyses of cohorts of twins from Sweden, Denmark, and Finland. N Engl J Med 343:78–85

Manolio TA, Bailey-Wilson JE, Collins FS 2006 Genes, environment and the value of prospective cohort studies. Nat Rev Genet 10:812–820

Moolgavkar SH, Stevens RG, Lee JA 1979 Effect of age on incidence of breast cancer in females. J Natl Cancer Inst 62:493–501

Parkin DM, Fernandez LM 2006 Use of statistics to assess the global burden of breast cancer. Breast J 12 Suppl 1:S70–80

Parkin DM, Bray F, Ferlay J, Pisani P 2005 Global cancer statistics, 2002. CA Cancer J Clin 55:74–108

Peto J, Collins N, Barfoot R et al 1999 Prevalence of BRCA1 and BRCA2 gene mutations in patients with early onset breast cancer. J Natl Cancer Inst 91:943–949

Risch N 2000 Searching for genetic determinants in the new Millennium. Nature 405:847–856

Shen YC, Chang CJ, Hsu C, Cheng CC, Chiu CF, Cheng AL 2005 Significant difference in the trends of female breast cancer incidence between Taiwanese and Caucasian Americans: implications from age-period-cohort analysis. Cancer Epidemiol Biomarkers Prev 14:1986–90

Sim X, Ali RA, Wedren S et al 2006 Ethnic differences in the time trend of female breast cancer incidence: Singapore, 1968–2002. BMC Cancer 6:261

Sinilnikova OM, McKay JD, Tavtigian SV et al 2007 Haplotype-based analysis of common variation in the acetyl-CoA carboxylase a gene and breast cancer risk: a case-control study nested within the European Prospective Investigation into Cancer and Nutrition. Cancer Epidemiol Biomarkers Prev 16:409–415

Sun CL, Yuan JM, Koh WP, Yu MC 2006 Green tea, black tea and breast cancer risk: a meta-analysis of epidemiological studies. Carcinogenesis 27:1310–1315

Vineis P, Schulte P, McMichael AJ 2001 Misconceptions about the use of genetic tests in populations. Lancet 357:709–712

Waard F 1979 Premenopausal and postmenopausal breast cancer: one disease or two? J Natl Cancer Inst 63:549–552

Ying CJ, Xu JW, Ikeda K, Takahashi K, Nara Y, Yamori Y 2003 Tea polyphenols regulate nicotinamide adenine dinucleotide phosphate oxidase subunit expression and ameliorate angiotensin II-induced hyperpermeability in endothelial cells. Hypertens Res 26:823–828

Yuan JM, Koh WP, Sun CL, Lee HP, Yu MC 2005 Green tea intake, ACE gene polymorphism and breast cancer risk among Chinese women in Singapore. Carcinogenesis 26:1389–1394

Ziegler RG, Hoover RN, Pike MC et al 1993 Migration patterns and breast cancer risk in Asian-American women. J Natl Cancer Inst 85:1819–1827

DISCUSSION

Koth: If you are using a big cohort and you are doing genome-wide association studies, shouldn't they be serving as each others' controls, in terms of variation, instead of the case control studies?

Chia: Within a cohort study, it is possible to nest a case-control study. As the incident cases appear, you take them and select a sample of controls from the cohort. The advantage in this is that you not only identify genetic but also environmental factors. Furthermore, the total number of cases and controls that you genotype will be far less than the entire cohort. In this sense, it is a case control study, but it is nested in a cohort design. One of the problems with a traditional case control study is that the controls may not be coming from the same study base, so it is not just a question of recall bias, but also a question of whether it is from the same study base.

Koth: In a sense they are not controls, it is just the variation in the population that you have. The other thing I wanted to ask is about other underlying factors. Are these also being documented in the study as covariates?

Chia: In most cohort studies you would be collecting quite a lot of exposure information. These data serve as covariates. For example, in our Singapore Chinese health study, apart from diet and physical activity, we have information on reproductive history, birth weight and so on.

Koth: Are traumatic events recorded?

Chia: The main outcomes of these cohort studies have been chronic diseases, such as cancer and cardiovascular disease, with less emphasis on behavioral and psychological endpoints. We have not collected information on traumatic events.

Braithwaite: In the postmenopausal Japanese women where the incidence tapers off, have you been able to stratify these populations relative to the Indians or Caucasians and do G × E or gene association studies?

Chia: Unfortunately, for that study it is based on cancer registration data without any biological specimens for genetic data.

Robertson: The ultimate susceptibility genes are *BRCA1* and *BRCA2*, which have a penetrance of less than 1, and at the other end of the spectrum is *CHEK2*, which has high prevalence but low penetrance. Have these been used as weather vanes to determine where the gold might lie in terms of environmental risk factors that might modify their penetrance? You could then reduce your power problem.

Chia: There is one study looking at *BRCA1* and family history which in a broad sense is an environmental factor. It doesn't modify the effect of *BRCA1*. There is a suggestion that perhaps parental age at first birth also modifies *BRCA1*.

Stankovich: I want to ask about the power for this sort of G × E genome-wide association study. You quoted a report saying that 20 000 cases would be needed. But in the context of G × G interactions an influential paper in *Nature Genetics* (Marchini et al 2005) argued that for certain models of G × G interaction there was comparable power with the same sample sizes that are required for detecting main gene effects. I am having trouble reconciling their conclusions with this conclusion that you need much larger sample sizes to detect G × E interactions.

Snieder: One issue is how many environmental factors you want to consider.

Stankovich: With G × G it is around 300 000 by 300 000 possible interactions!

Snieder: That is true.

Martin: A few weeks ago one of my ex-students presented a paper at a conference in which he did a G × G interaction analysis for the genome-wide association for bipolar disorder. I think he had something like 49 billion possible interactions, so you can imagine the type I error problem!

Stankovich: This *Nature Genetics* paper was arguing that you first test for main effects, then set a low threshold for follow-up testing for G × G interactions.

Chia: I suppose one of the reasons why G × E studies need to be big is that the environmental variation is much larger. Take as an example a dietary questionnaire: there are many steps to get down to the nutrients, and there is a lot of noise.

Rutter: I am intrigued by your migration data. The migration research strategy seems to me to be much under-utilized. The differences you were demonstrating were really quite striking. What are the genetic differences between the groups that might be relevant to the initial finding in the original country? And do we know what the lifestyle differences are that make the risks go up in line with the host country?

Chia: In terms of genetic differences, we don't have those data. In terms of lifestyle, it is the effect of a westernized lifestyle and it takes about two generations before the effect comes in.

Braithwaite: There is evidence from some Australian studies (and elsewhere) that European migrants have lower incidence of colorectal cancer, but after time in Australia, they reach the same level of cancer incidence as the locals (Australians) (McMichael et al 1980, Grulich et al 1995, Tyczyński et al 1994).

Rutter: That is intriguing but puzzling. Is this just an environmental effect or does it represent some kind of gene–environment interaction? The data suggest the latter, but haven't demonstrated it as yet.

Braithwaite: The authors of some of the work mentioned above argue for a high association with the intake of red meat, which is higher in Australians than Europeans, but changes in the migrants. But the G × E hasn't been done to my knowledge.

Martinez: There is a very interesting example with asthma with immigrants from Mexico who arrive in the USA at different ages. If you arrive in the USA when you are young you have almost the same prevalence of asthma as an adult as those who are born in the USA and who are not Mexican. But if you arrive at older ages you have less asthma. If you arrive at the age of 20 you have the same asthma risk as those born in Mexico (Eldeirawi et al 2005).

Kotb: This is extremely interesting. There is a relationship between depression and the immune system. This especially applies to natural killer (NK) cells, which are the main cells that fight cancers. There are also data that in depressed people that NK cells are down by 80%. So depression can certainly influence cancer. Coming back to the context of migration studies, Mediterranean people tend to party a lot, socialize and laugh a lot, whereas when they emigrate they tend to have heavier work loads and do little socialization. These social differences could affect emotions, depression and in turn the immune system. Is this another G × E interaction?

Martin: I don't want to sound like a cracked record, but can I make another plug for how useful MZ twins can be for this sort of work? The MZ twinning rates are constant around the world at 4 per 1000, so there must be huge samples of MZ twins available if you try hard enough.

Rutter: That doesn't help much with migration studies: the sample size of MZ twins, one of whom has migrated and the other of whom hasn't must be tiny.

Martinez: Is the penetrance of these genes the same for different ethnic groups? In other words, is the penetrance of *BRCA1* less for the Chinese?

Chia: They are extremely rare, even in the Asian population. From the small samples we have, the penetrance seems the same.

Snieder: One of the things that we have been discussing is that we need better measures of the environment. You mentioned that you will be collecting food frequency questionnaire data. What is the value of this? Measuring diets accurately is extremely difficult.

Chia: There is a lot of debate about the best way of collecting dietary data, especially with large populations. Food frequency questionnaires came out on top, primarily because they are logistically easier to implement. The ideal would be to have a 24 h recall, but data capture is the problem here. We are coming up with an IT tool to try to capture 24 h recall that will be connected to a database. If the food item is new, it can be given an entry. We can even collect recipes of that item to come up with the nutrients. This means that we must have a nutritional database in the background as the engine, and this is quite tough to come up with for different populations. For the Singapore Chinese health study it took us about three years to devise this database. To extend this to the Malay diet and the Indian diet is complicated.

Martin: There has been a huge amount invested in trying to get dietary information in relation to breast cancer. What has come out of it? The risk factors associated with diet are trivial compared with reproductive risk factors. Is it really worth doing this expensive work?

Chia: If you are looking at cancer as the outcome, I am not sure that dietary information is that useful. But it has been shown to be useful for cardiovascular outcomes. We are primarily looking at metabolic disease outcome and cardiovascular disease outcome.

Kotb: Instead of looking at gene interactions, if you cluster the genes into pathways and then give it a score based on any polymorphism in each specific pathway that would have a functional consequence, would this bring up further interactions that are more realistic from a physiological point of view? And perhaps would this help generate some hypotheses that can be experimentally tested?

Chia: Theoretically, if we use the pathway approach it could make more sense. But we did a similar kind of study with Swedish samples looking at estrogen metabolism pathways, and identified a total of about 1600 SNPs along these pathways. We came up with nothing. Perhaps we don't know the pathways well enough to be able to identify suitable markers. With the resolution with which we are able to do genetic markers, we have to do this fishing expedition rather than a pathway or candidate gene approach.

Snieder: What about the new genes that have been thrown up by the genome-wide association? Is there anything known about their potential function?

Chia: For these four genes, three are known, but one is totally unknown.

Snieder: This is part of the excitement, to find new mechanisms and insights.

Braithwaite: A common thing from cell culture studies is that if breast epithelial cells are cultured they almost always grow out after a few weeks. Always, with no exception, they silence the p16 gene though mutation, deletion or methylation of the promoter sequences. Has anyone tried to look to see whether there are any silencing effects in a whole range of genes?

Chia: I am not familiar with this area.

Reeve: Did you say that in the Chinese population, irrespective of the McDonalds diet, postmenopausally the risk of breast cancer is flattening off, so there is a clear difference between the two populations?

Chia: Yes.

Reeve: You are talking about accumulating 20000 people from the population. Only a small proportion of these would have tumors. If you compared gene expression profiles of the Chinese before and after menopause, you would be able to use pathway analysis to get some sort of clue of what the environmental factors might be. You could compare it with an equivalent North American cohort.

Chia: We have been doing expression profile from fresh frozen tissue to compare the Chinese and the Swedish populations.

Reeve: The key thing here is to compare post and pre-menopause. There is something strange occurring in the Chinese population where it flattens off.

Chia: One difference is the estrogen receptor (ER) status. In Caucasian populations it is around 90%, whereas in Asian populations it is about 50%.

Braithwaite: That high ER status might be important. There is an MDM2 SNP that discriminates breast cancer incidence in premenopausal from postmenopausal women as expression of MDM2 becomes responsive to estrogen. This in turn down-regulates p53. If this status differs in the Asians, it might account for the difference in the shapes of the curves.

Chia: The risk factors are different. For the Caucasian population it is driven primarily by hormone-replacement therapy (HRT), but in Asian populations which use very little HRT it is driven primarily by reproductive factors.

Reeve: The histopathology could be important. I have had the same experience myself comparing the histopathology of an embryonal tumor between Japanese and North Americans and white New Zealanders. It is only when I got a top class histopathologist looking at this that we could pick up the differences. This is what you need to do.

Chia: We have a molecular biologist looking at tissue blots of the collections we have, from premenopausal breast cancer and postmenopausal breast cancer from a number of ethnic groups.

Rutter: Can you say more about the potential of intermediate endpoints. Yesterday we had some discussion of endophenotypes. The point you are making here is an equally important but different one. If one is looking at something that takes a long time to develop, to wait for all that to happen is pretty hopeless. How good are the available endpoints of an intermediate kind? You mention mammographic density. How good is that?

Chia: So far for breast cancer, mammographic density is the main intermediate endpoint. Some people have tried to do lavage from the breast, but this is quite invasive. The problem of mammographic density is quantifying it in a standard fashion.

Battaglia: Radiologists have been debating for years about automatic versus human-operated rating of radiographies and diagnosis. That's perhaps why they were prompted to invent ROC analysis, I guess. The discussion is ongoing.

Chia: The question is, is it for clinical use or for research use? Right now we have 33 000 mammograms!

Rutter: The measurement issues are horrendous. The need for very large samples is undeniable for some sorts of things, but the price you pay in terms of measurement is awesome. We need big samples, but unless we have really good measures we are equally lost.

References

Eldeirawi K, McConnell R, Freels S, Persky VW 2005 Associations of place of birth with asthma and wheezing in Mexican American children. J Allergy Clin Immunol 116:42–48

Grulich AE, McCredie M, Coates M 1995 Cancer incidence in Asian migrants to New South Wales, Australia. Br J Cancer 71:400–408

Marchini J, Donnelly P, Cardon LR 2005 Genome-wide strategies for detecting multiple loci that incluence complex diseases. Nat Genet 37:413–417

McMichael AJ, McCall MG, Hartshorne JM, Woodings TL 1980 Patterns of gastro-intestinal cancer in European migrants to Australia: the role of dietary change. Int J Cancer 25:431–437

Tyczyński J, Tarkowski W, Parkin DM, Zatoński W 1994 Cancer mortality among Polish migrants to Australia. Eur J Cancer 30A:478–584

10. Unbiased forward genetics and systems biology approaches to understanding how gene–environment interactions work to predict susceptibility and outcomes of infections

Malak Kotb*†‡, Nourtan Fathey*†‡, Ramy Aziz*†‡, Sarah Rowe†‡, Robert W. Williams† and Lu Lu†

*Department of Molecular Genetics, Biochemistry and Microbiology, The University of Cincinnati, Ohio, †The MidSouth Center for Biodefense and Security at The University of Tennessee Health Sciences Center and ‡The VA Medical Center, Memphis TN 38163, USA

Abstract. Like most human diseases, infectious diseases are effected by complex genetic traits and multiple, interactive environmental and inherent host factors. By linking specific genotypes to disease susceptibility phenotypes we can identify the genetic basis for inter-individual differences in disease susceptibility as well as gain insight into how gene–environment interactions influence infection outcomes. Our research has focused on delineating interactive pathways and molecular events modulating host resistance or susceptibility to specific pathogens. Our model system has been that of Group A *Streptococcus* infections that can manifest in starkly different ways and cause distinct diseases in genetically distinct individuals. We have extended our work to other pathogens, including those with a potential of causing major, global biological threats. In as much as it is quite difficult to conduct certain infectious disease studies in humans, there has been a critical need for small animal models for infectious diseases. Appreciating the limitations of existing models, we developed several novel and complementary mouse models that are ideal for use in systems genetics studies of complex diseases. These models not only allow biological validation of known genetic associations, but importantly they afford an unbiased tool for discovering novel genes and pathways contributing to disease outcomes, under different environments.

2008 Genetic effects on environmental vulnerability to disease. Wiley, Chichester (Novartis Foundation Symposium) p 156–167

Like most human diseases, infectious diseases have important genetic and environmental components (Kotb 2004). Indeed, host genetic variability and complex polymorphic and non-polymorphic interactive traits affect susceptibility to infections and strongly potentiate the severity of clinical manifestations associated with it. In an effort to better understand how gene–environment interactions work to influence susceptibility and outcomes of specific infections, our approach has been to compare the genetic makeup and immune responses of susceptible and resistant individuals and to link specific host genotypes to disease susceptibility and phenotypes to identify genes and pathways involved in potentiating infection outcomes. In as much as all infectious diseases have an important immunological component, polymorphic genes encoding immune effector mediators or regulators are the most likely candidates to impact infection susceptibility. Examples include genes encoding cytokines, adhesion molecules, chemokines, chemokine receptors, homing molecules, molecules involved in cell–cell interactions, signaling molecules and so on. We have taken an unbiased systems genetics approach to discover genes and pathways that play an important role in specific infections, and our long-term goal is identify those that are common for several infections and those that are unique to specific ones.

Our model pathogen has been Group A *Streptococcus* bacteria (*S. pyogenes*), which are interesting bacteria that can cause a wide variety of human diseases depending, to a great extent, on the genetic make up of the infected host as well as environmental factors. Diseases caused by these bacteria range from a simple uncomplicated sore throat (pharyngitis) to severe and life-threatening illnesses. The bacteria normally reside in the throat or on the skin but occasionally they invade their way into host sterile sites. Several weeks post-infection, some patients develop non-suppurative acute rheumatic fever (ARF), and while most recover with no complications others suffer chronic and debilitating autoimmune sequelae such as rheumatic carditis and rheumatic heart disease (RHD), glomerulonephritis (GN) and Sydenham's Chorea. Invasive diseases caused by these bacteria can manifest as either mild cellulitis or as very severe and fatal illnesses such as streptococcal toxic shock syndrome (STSS) and necrotizing fasciitis (NF) (Cunningham 2000).

Although there are over 100 different serotypes of these bacteria, the same strain can be isolated from asymptomatic individuals as well as from patients with starkly different clinical manifestations. This clearly underscores the important role played by the host in these infections. Indeed, we have been able to demonstrate associations between polymorphic MHC genes and different manifestations of Group A *Streptococcus* bacteria (Norrby-Teglund et al 2000, Norrby-Teglund & Kotb 2000, Kotb et al 2002), and these associations have been biologically validated *in vitro* using human cell cultures, and *in vivo* using a variety of animal models including transgenic mice expressing the polymorphic human MHC genes of interest. To illustrate our approach to identifying the genetic basis of infectious diseases, we will present an example from our studies of patients with invasive Group A streptococcal infections (mostly GAS sepsis patients) who presented with either severe sepsis and STSS or with uncomplicated, mild bacteraemia. This marked individual variation in invasive disease severity lead us to ask what might be the biological processes involved in disease progression and which polymorphic genes may be affecting it.

GAS elaborate a vast repertoire of virulence factors that can be functionally subclassified into mediators of bacterial adhesion and invasion, evasion of host defenses, inflammation or toxic destruction; but some of them may have overlapping roles in modulating host defenses (Cunningham 2000, Norrby-Teglund & Kotb 2000, Bisno et al 2003, Kotb et al 2003, Kreikemeyer et al 2003, Chhatwal & McMillan 2005). The relevant importance of these virulence factors to the various GAS diseases depends on where and when they are produced by the bacteria as well as on the host response to them. In the case of streptococcal sepsis and STSS, the pathogenesis is driven primarily by a group of immune stimulators known as superantigens (SAgs) (Kotb 1995, 1998), which interact simultaneously with resting T cells and MHC II-expressing cells, triggering their excessive activation and the massive release of inflammatory cytokines (e.g. TNFβ and IFNγ). Uncontrolled, overproduction of these inflammatory cytokines mediates many of the pathology seen in sepsis, leading to organ failure, capillary leakage and death (Norrby-Teglund & Kotb 2000, Kotb et al 2003, Norrby-Teglund et al 2003).

GAS are the richest bacteria with respect to their SAgs repertoire with at least 12 different SAgs—one of which has at least 10 different alleles (Proft & Fraser 2007), and most strains produce 3–4 SAgs. It was perplexing, therefore, the fact that patients with SAg producing GAS sepsis present with starkly different manifestations and severity. In fact, only few septic individuals develop the most severe form of GAS sepsis, STSS. The reason for this, we found, is that immune responses of different individuals to the same SAg can vary quite significantly (Kotb 1995, Kotb et al 2002, 2003, Nooh et al 2007). This prompted us to think which host factor(s) are likely to modulate the magnitude of the inflammatory response to the SAgs, and although we were faced with potential candidates, the most likely one was the MHC class II molecules, which serve as SAg receptors and are highly polymorphic (Kotb 1992). In fact, we and others had shown that allelic polymorphisms in HLA class II molecules affect SAg responses *in vitro* (Al-Daccak et al 1994, 1998, Norrby-Teglund et al 2002). Accordingly, we launched an epidemiological study of the HLA association with invasive Strep infection outcomes and were able to identify haplotypes that confer either protective or high risk in severe disease and were able to provide biological validation for the direct role of HLA allelic polymorphism using human cells and HLA transgenic mouse models as detailed below (Kotb et al 2002, 2003, Nooh et al 2007).

Considering the complex nature of sepsis in general and GAS sepsis in particular we took systems genetics approach to identify complex host genetic polymorphisms and interactive pathways affecting the overall outcome of this disease.

Results and discussion

Although the severe invasive GAS infections had subsided after the 1920s pandemics, a striking resurgence in the most severe forms of those diseases took place in the early 1980s with a remarkable increase in the incidence of streptococcal toxic shock (STSS) and necrotizing fasciitis (NF) reported particularly in Canada, the USA and Scandinavian countries (Low et al 1998). Our study focused on patients from Ontario, Canada where disease incidence was higher than other parts of the world and the CDC, together with infectious disease physicians in Mt. Sinai Hospital in Toronto had launched an active surveillance study in all of the Ontario area where 150 hospital and laboratories reported every case to our collaborators Drs Donald E. Low and Allison McGeer. Prior to initiating the study a group consisting of basic and clinical scientists that included statisticians and epidemiologists spent considerable effort on the design of the patient clinical database to make sure that all relevant information are collected including the patient's clinical history. Case definition and exclusion criteria were clearly defined, and protocols for sample collection and processing were standardized.

In the first phase of our studies we assessed differences in protective humoral immunity to the particular bacterial strain recovered from each patient and found that low levels of antibodies that aid in the clearance of the bacteria by phagocytic cells (i.e. serotype-specific anti-M protein antibodies) as well as lack of SAg-neutralizing antibodies increased the risk of the bacteria accessing the patient's sterile sites to initiate invasive disease (Basma et al 1999). However, the low levels of these protective antibodies did not correlate with disease severity. Instead, there was a marked increase in the production of inflammatory cytokines in patients with severe manifestations and STSS as compared to those with mild bacteraemia (Norrby-Teglund et al 2000). This difference in systemic inflammatory responses to the SAgs was reproducible even after their recovery, where those who experienced severe sepsis continued to mount high responses to the SAgs produced by their infecting strain, whereas those who had mild bacteraemia maintained a low level of response. This stable propensity to be a high or low responder to the Strep SAgs suggested an underlining genetic basis.

We set out to elucidate the genetic and molecular basis for the stark difference in GAS sepsis severity among our patient cohort, and we focused on genetic elements that may potentiate the host response to GAS SAgs. Confounding factors, such as age and underlying disease that may lead to misclassification were addressed in the final analysis. Our analysis focused on HLA class II polymorphisms and we identified specific HLA-II alleles and haplotypes that confer strong resistance to

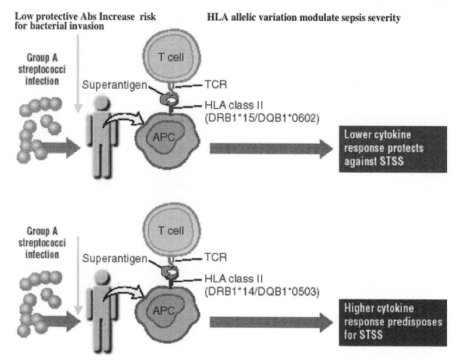

Low protective Abs Increase risk   HLA allelic variation modulate sepsis severity
for bacterial invasion

FIG. 1.   Environmental and genetic factors modulate susceptibility to severe streptococcal
sepsis. (Modified from Kotb 2004.)

STSS, NF or a combination of both in the same patient and others that predispose
to these illnesses (Kotb et al 2002) (Fig. 1). The DRB1*1501-DQB1*0602 haplotype conferred strong protection against STSS, while the DRB1*14-DQB1*0503
haplotype increased the risk for it. Interestingly, HLA associations with NF were
distinct from those with STSS. Among those with soft tissue infections, the
DRB1*03-DQB1*0201 haplotype was significantly associated with protection
from NF, and the DRB1*11-DQB1*0301 haplotype showed a trend toward an
association with risk for NF among invasive cases without STSS compared with
healthy controls. The presence of the DRB1*1501-DQB1*0602 haplotype in NF
cases protected them from developing a combination of STSS plus NF, the deadli-
est form of invasive disease (Kotb et al 2002, 2003).

We validated the direct role of HLA-II association, biologically, through both
in vitro studies with human PBMC expressing different HLA types as well as in
vivo studies with HLA-tg mice carrying alleles of interest (Nooh et al 2007) (Fig.
2). Both recovered patients and healthy subjects who carried the protective
DRB1*1501-DQB1*0602 haplotype consistently mounted significantly lower

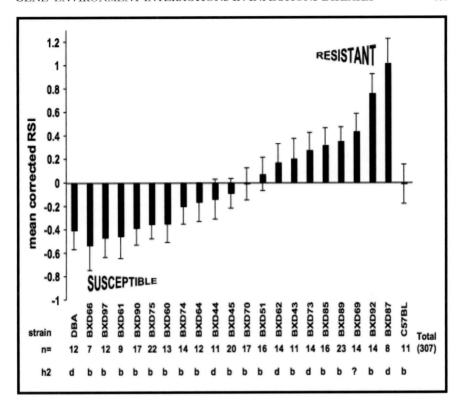

FIG. 2.   Genetic variation and complex traits modulate severity of GAS sepsis in ARI BXD mice and reveal more exaggerated or suppressed phenotypes than the parental strains. The bar chart shows the mean values of corrected relative survival indices (cRSI) for 20 BXD strains, arranged in ascending order. Parental strains (C57Bl/6J and DBA/2J) are shown on the two extremities of the x axis. Error bars represent the standard errors of the means. The total number of animals (*n*) used per strain is indicated. (Data from Aziz et al 2007.)

responses than those who lack this haplotype, and HLA-tg mice expressing either allele on the protective haplotype were resistant to severe GAS sepsis and survived the intravenous infection, whereas mice expressing neutral HLA-II alleles succumbed to it.

As mentioned above, that HLA-II polymorphisms contribute to STSS susceptibility is logical because Strep SAgs, which are pivotal mediators of STSS, utilize the HLA-II molecules as receptors through which they interact with TCRVβ elements and elicit potent inflammatory responses leading to STSS in genetically predisposed, high responders (Fig. 1). But how about those who neither carry the protective nor the high risk HLA-II haplotypes or alleles? What other genetic factors might have contributed to their disease outcomes and what is the role of additional virulence factors produced by the bacteria in the disease process? To address these questions we needed to consider the overall complex genetic susceptibility to the different manifestations of GAS infection.

We reasoned that a systems genetics approach may be the ideal route to address the above questions and to discover additional genetic variations and pathways that modulate the outcome of GAS sepsis. Accordingly, we turned to the advanced recombinant inbred (ARI) BXD mouse panel derived from the parental C57BL/6J and DBA/2 strains, known to differ considerably in their susceptibility to a number of infectious agents.

The advanced recombinant inbred are very useful in systems genetics studies, because these mice represent a genetically diverse population of homozygous, inbred lines, each of which is genetically distinct (Williams et al 2001, Wang et al 2003, Peirce et al 2004, Chesler et al 2005, Aziz et al 2007). Genetic differences among the various BXD strains are easily trackable because the parental strains are fully sequenced and each inbred strain is heavily genotyped with more than 3600 genomic markers. Thus by linking differences in disease phenotyes across the panel of the BXD strains to the genotype of each strain, identification of disease quantitative trait loci (QTLs) is relatively straightforward and achievable in a short period of time (Williams et al 2001, Wang et al 2003, Peirce et al 2004, Chesler et al 2005, Aziz et al 2007).

We compared susceptibility to severe GAS sepsis in over 40 BXD strains, each represented by a minimum of five mice and an average of 15 mice per strain. Because each BXD strain is represented by inbred mice, it was possible to combine data from experiments performed months apart, with highly reproducible results. Mice (ages 40–120 days) from various BXD strains were infected via the tail vein with $1–3 \times 10^7$ CFU of the virulent M1T1 GAS clinical isolate in 100 µl saline. We monitored differences in three phenotypes across the panel: survival, weight loss and bacteraemia. Bacterial load in blood (CFU/ml) was determined for all mice at 24 h, and a bacteraemia index was determined and corrected for covariates. All mice developed bacteraemia, but there were significant differences across the strains and the same was true for infection severity and survival rates. Mouse survival was monitored every 8 h for 7 d. To normalize across experiments, we inspected survival days distribution clusters for each experiment and determined multimodal distribution and boundaries of each cluster for a total of three clusters: susceptible, intermediate, and resistant. Survival days within each cluster were then converted into a survival index ranging from 0.25–1, 1.25–2, and 2.25–3 for susceptible, intermediate, and resistant clusters respectively. The survival index was assigned to each mouse irrespective of its strain. Indices for each strain, across experiments were then corrected for significant covariates (age, sex, body weight, and inocula) using multiple regression analyses. Mice were also weighed every 12 h to monitor weight loss for each infected mouse (Abdeltawab et al 2008).

As shown in Fig. 2, several BXD strains exhibited phenotypes outside the ranges of the parental strains, with several significantly more susceptible or resistant than their ancestors. Age was confirmed as a significant determinant of survival and bacterial spread, but the strongest factor influencing survival, as expected, was the genetic background of BXD strains ($P \leq 0.0001$). We analyzed differences in disease phenotypes across the BXD strains in the context of each strain genotype using WebQTL tools available on the GenNetwork site (*http://www.genenetwork. org*), and this analysis revealed a strong QTL modulating sepsis severity and mouse survival to two loci on chromosome 2. The strongest QTL mapped to mouse Chr 2 between 22–34 Mb, with an likelihood of the odds (LRS) of 34.2 ($P \leq 0.0000001$), and the second less significant QTL on the same chromosome was between 125–150 Mb with an LRS 130 of 12 ($P \leq 0.001$) (Abdeltawab et al 2008). A third QTL was mapped to the X chromosome. Using haplotype maps for the BXD panel we selected additional strains that allowed us to narrow down the mapped interval.

The narrowed down QTLs were found to harbor several polymorphic genes known to regulate immune responses to bacterial infections. We evaluated candidate genes within this QTL using multiple parameters that included linkage, gene ontology, variation in gene expression, cocitation networks, and biological relevance, and identified pathways involving *Il1a* that encodes the inflammatory cytokine IL1a, and *Ptges* that encodes prostaglandin E synthase as key networks involved in modulating GAS sepsis severity (Abdeltawab et al 2008). The association of GAS sepsis with multiple pathways underscores the complexity of traits modulating GAS sepsis and provides a powerful approach for analyzing interactive traits affecting outcomes of other infectious diseases.

We believe the above studies illustrate a systematic approach to dissecting gene × environment interactions that affect susceptibility and outcomes to infectious diseases. A combination of targeted and systems genetic analyses allowed us to identify various mechanisms and molecular events involved in modulating host–pathogen interactions and affecting disease outcomes. Coupled with the ability to biologically validate the data in various *in vitro* and *in vivo* setups, using appropriate models for each, one can extract valuable data even for relatively rare infections where the number of cases available for study may not reach the needed statistical power. Ultimately the goal is to translate the information in collaboration with physicians to develop sensitive and specific diagnostics and more effective and targeted interventions and therapeutics for these infections.

Acknowledgements

The work presented was supported by funds from grant AI4 0198-06 from The National Institute of Allergy and Infectious Diseases NIAID (M.K.); Merit Award from the Research and Development Office, Medical Research 404 Service, Department of Veterans Affairs (M.K.); and the U.S. Army Medical Research Acquisition Activity Grant W81XWH-05-1-0227 (M.K.).

References

Abdeltawab NF, Aziz RK, Kansal R et al 2008 An unbiased systems genetics approach to mapping genetic loci modulating susceptibility to severe Streptococcal sepsis. Submitted

Al-Daccak R, Mehindate K, Poubelle PE, Mourad W 1994 Signaling via MHC class II molecules selectively induces IL-1 beta over IL-1 receptor antagonist gene expression. Biochem Biophys Res Commun 201:855–860

Al-Daccak R, Mehindate K, Damdoumi F et al 1998 Staphylococcal enterotoxin D is a promiscuous superantigen offering multiple modes of interactions with the MHC class II receptors. J Immunol 160:225–232

Aziz RK, Kansal R, Abdeltawab NF et al 2007 Susceptibility to severe Streptococcal sepsis: use of a large set of isogenic mouse lines to study genetic and environmental factors. Genes Immun 8:404–415

Basma H, Norrby-Teglund A, Guedez Y et al 1999 Risk factors in the pathogenesis of invasive group A streptococcal infections: role of protective humoral immunity. Infect Immun 67:1871–1877

Bisno AL, Brito MO, Collins CM 2003 Molecular basis of group A streptococcal virulence. Lancet Infect Dis 3:191–200

Chesler EJ, Lu L, Shou S 2005 Complex trait analysis of gene expression uncovers polygenic and pleiotropic networks that modulate nervous system function. Nat Genet 37:233–242

Chhatwal GS, McMillan DJ 2005 Uncovering the mysteries of invasive streptococcal diseases. Trends Mol Med 11:152–155

Cunningham MW 2000 Pathogenesis of group A streptococcal infections. Clin Microbiol Rev 13:470–511

Kotb M 1992 Role of superantigens in the pathogenesis of infectious diseases and their sequelae. Curr Opin Infect Dis 5:364

Kotb M 1995 Bacterial exotoxins as superantigens. Clin Microbiol Rev 8:411–426

Kotb M 1998 Superantigens of gram-positive bacteria: structure-function analyses and their implications for biological activity. Curr Opin Microbiol 1:56–65

Kotb M 2004 Genetics of susceptibility to infectious diseases. ASM News 70:457–463

Kotb M, Norrby-Teglund A, McGeer A et al 2002 An immunogenetic and molecular basis for differences in outcomes of invasive group A streptococcal infections. Nat Med 8:1398–1404

Kotb M, Norrby-Teglund A, McGeer A, Green K, Low DE 2003 Association of human leukocyte antigen with outcomes of infectious diseases: the streptococcal experience. Scand J Infect Dis 35:665–669

Kreikemeyer B, McIver KS, Podbielski A 2003 Virulence factor regulation and regulatory networks in Streptococcus pyogenes and their impact on pathogen-host interactions. Trends Microbiol 11:224–232

Low DE, Schwartz B, McGeer A 1998 The reemergence of severe group A streptococcal disease: an evolutionary perspective. In: Scheld WM, Armstrong D. Hughes JM (ed) Emerging infections. ASM Press, Washington DC, p 93–123

Nooh MM, El-Gengehi N, Kansal R, David CS, Kotb M 2007 HLA transgenic mice provide evidence for a direct and dominant role of HLA class II variation in modulating the severity of streptococcal sepsis. J Immunol 178:3076–3083

Norrby-Teglund A, Kotb M 2000 Host-microbe interactions in the pathogenesis of invasive group A streptococcal infections. J Med Microbiol 49:849–852

Norrby-Teglund A, Chatellier S, Low DE, McGeer A, Green K, Kotb M 2000 Host variation in cytokine responses to superantigens determine the severity of invasive group A streptococcal infection. Eur J Immunol 30:3247–3255

Norrby-Teglund A, McGeer A, Kotb M et al 2003 Severe invasive group A Streptococcal infections. Kluwer/Plenum, New York

Norrby-Teglund A, Nepom GT, Kotb M 2002 Differential presentation of group A streptococcal superantigens by HLA class II DQ and DR alleles. Eur J Immunol 32:2570–2577

Peirce JL, Lu L, Gu J, Silver LM, Williams RW 2004 A new set of BXD recombinant inbred lines from advanced intercross populations in mice. BMC Genet 5:7

Proft T, Fraser JD 2007 Streptococcal superantigens. Chem Immunol Allergy 93:1–23

Wang J, Williams RW, Manly KF 2003 WebQTL: web-based complex trait analysis Neuroinformatics 1:299-308

Williams RW, Gu J, Qi S, Lu L 2001 The genetic structure of recombinant inbred mice: high-resolution consensus maps for complex trait analysis Genome Biol 2:RESEARCH0046

DISCUSSION

Braithwaite: That's fantastic, but it would probably cost the entire health research budget of New Zealand to do it! In order to test the genetic involvement in a variety of genetic diseases, those disease pathogens have to be able to infect mice. There are many viruses that probably don't, or if they do, they might bind to the surface of cells and not go through a lytic cycle. Do you need to use pathogens that replicate the full lytic cycle seen in human contexts?

Kotb: Yes and no. In many cases where people who say the mouse isn't a good model for their pathogen, when they start to use this genetically diverse population they see phenotypes that mimic some or many of what is seen in the human population. But this doesn't mean that every pathogen will work in these mice. So far we are hitting on a much larger repertoire of pathogens that are working well in these panels. When we get the 1000 crosses, we expect to see even more phenotypes that reflect human diseases. The different combination and context of polymorphic traits affects the disease phenotype. We see phenotypes that are either grossly exaggerated or completely suppressed in the progeny of the BXD mice compared to their parents B6 and D2 mice. This is a manifestation of how phenotypes can differ drastically depending on the complex genetic make up and genetic context of the host. The nice thing is that you can reproduce interesting differences in phenotypes in different BXD lines as many times as you wish—it is like doing studies with outbred mice except the results are reproducible.

Braithwaite: Some pathogens will replicate well in their natural hosts and yet don't replicate at all in unnatural hosts. But you indicated that you might be able to determine permissivity factors by using mice of this variety, or cells derived from them, to look for factors that control virus permissivity or tropism. That is, genetic factors that affect the species tropism of the pathogens.

Kotb: Exactly. And in few cases when we do not see differences in phenotypes in these mice we can envisage introducing the polymorphic human gene that gives us this difference, as we did with the HLA transgenic mice. This is why I added these mice to the presentation. The mice normally respond poorly to the strep superantigens, but when different HLA class II alleles are introduced into them you get the susceptibility and protected phenotypes that mimic what we see in humans. We are doing transcriptome analysis on these mice in the resting state in different organs, and analyzing the data to look at co-regulated genes. We are subtyping based on genes linked by common circuitries. If you perturb the common regulator you are likely to perturb these genes together. They move as blocks together. We are doing this under normal conditions in different cells and then under different disease conditions. The normal data will be placed on the website for free access to the public, so that other investigators who would like to do their own perturbations can have a reference to compare with.

Robertson: I have a comment about how you infect your mice. Many of these pathogens won't naturally adopt a percutaneous route of infection in humans. Is there any thought of mimicking normal infection routes, and in so doing examining mucosal barrier functions?

Kotb: We have different models: subcutaneous, intranasal and intratracheal. Standardizing the dose for each route is also important. We are trying to reduce the variables as much as possible. We are looking at variation in disease phenotypes as a result of genetic variation, and more importantly the combination of the genetic variation. To determine how certain polymorphic genes behave when they are in the context of other polymorphic genes.

Braithwaite: Could you use this mouse library to look for modulating alleles for cancer susceptibility? My problem with the experiment might be that if you start off with an inbred mouse which has an oncogene on it, and cross it onto a series of mice which you now have homozygosity for, does the background of the crossed mouse cause a problem in looking for the predisposing alleles?

Kotb: Yes you can use these mice to look for modulating alleles for cancer susceptibility. If you cross a mouse that has the oncogene onto the BXD mice you will see resistant, susceptible and in between phenotypes because the phenotypic expression of this gene will be affected by the genetic context of each strain you will be crossing your mouse onto. This is again why in most cases we do not see 100% association between gene and disease. It is like the *BCRA1* and *BCRA2* genes and susceptibility to breast and ovarian cancer. Not everyone who has the predisposing mutations in these genes has the disease and vice versa. This suggests that the effect of the mutation likely depends on the genetic context of the host as well as environmental factors. The recombinant inbred mice were originally generated for use in cancer studies and the effect of differences in overall genetic context is being studied. What other things can exaggerate the effects of a disease association gene, if it really is a major gene for disease pathology, or are you going to see suppression? Or are you going to see something else that may have been more important. The kind of data you get from these analyses is similar to what you get from transcriptome analysis: you get a lot of data, and you are putting those differentially expressed genes into pathways, and this tells you where you should focus and test your hypotheses in an unbiased way. The eight way cross that is underway to generate the 1000 crosses will help get us closer to human diseases and phenotypes, when we introduce the gene of interest and looking at how this gene now behaves in the context of much higher level of genetic variation.

Tesson: I have a basic question. I do understand how these mice can be useful for looking at different diseases, but I don't understand how you can increase the diversity. You have only eight strains at the beginning, and these were inbred strains to start with. Yes, there are thousands of combinations, but there are only eight different possible alleles.

Koth: The mice are bred in such a way to maximize recombination within each chromosome. This generates tremendous variation and then the polymorphisms resulting form these recombinations are in the context of different sets of other similarly polymorphic alleles as a result of crossbreeding, and then there is more recombination as you crossbreed before you start inbreeding selected pairs to be the parents of each inbred line. Remember, we all came from the same parents, Adam and Eve, and look at us now.

Tesson: Imagine a SNP associated with a disease in a given population, if this SNP was not present in one of the eight original strains, then you are not going to find it.

Koth: If it is something that is so precise that you need that one single SNP by itself to cause the disease, yes we may miss it. But if it is a disease that is affected by complex traits, the likelihood is that you will see it.

11. Gene–environment interactions in environmental lung diseases

Steven R. Kleeberger and Hye-Youn Cho

Laboratory of Respiratory Biology, National Institute of Environmental Health Sciences, National Institutes of Health, Research Triangle Park, North Carolina, USA

Abstract. Lung diseases such as asthma, chronic obstructive pulmonary disease (COPD), and acute respiratory distress syndrome (ARDS) have complex etiologies. It is generally agreed that genetic background has an important role in susceptibility to these diseases, and the genetic contribution to disease phenotypes varies between populations. Linkage analyses have identified some predisposing genes. However, genetic background cannot account for all of the inter-individual variation in disease susceptibility. Interaction between genetic background and exposures to environmental stimuli, and understanding of the mechanisms through which environmental exposure interact with susceptibility genes, is critical to disease prevention. Use of animal models, particularly inbred mice, has provided important insight to understand human disease etiologies because genetic background and environmental exposures can be controlled. We have utilized a positional cloning approach in inbred mice to identify candidate susceptibility genes for oxidant-induced lung injury. Subsequent investigations with cell models identified functional polymorphisms in human homologues that confer enhanced risk of lung injury in humans. This 'bench to bedside' approach may provide an understanding of gene–environment interactions in complex lung diseases is essential to the development of new strategies for lung disease prevention and treatment.

2008 Genetic effects on environmental vulnerability to disease. Wiley, Chichester (Novartis Foundation Symposium) p 168–180

Over the last several decades the incidence of environmental diseases such as asthma has increased with alarming frequency in industrialized cities worldwide (e.g. Elias et al 2003). These diseases generally are complex, with clear contributions of genetic background and exposure to environmental stimuli (see Kleeberger & Peden 2005). It is unlikely that the increased incidence in disease can be attributed only to genetics as increases in disease-causing genetic mutations to account for the increase would require multiple generations. Therefore the role of environmental exposures and stimuli has received increasing attention as one of the causative factors in increased disease prevalence. However, it is likely that neither genetics nor environment alone cannot account for the disease increase, but the interaction of these two factors (i.e. gene × environment) must be considered.

A genetic contribution to many or most diseases has been demonstrated. The recent emergence of genotyping capabilities and technologies, as well as the continued refinement of the human genome sequence has facilitated identification of genetic underpinning of human diseases. Candidate genes studies have provided useful information on the role of specific genes. Candidate genes are generally chosen for investigation in human populations based on biological and mechanistic plausibility derived from cell and animal models. Family-based linkage analyses have also identified candidate susceptibility genes. These investigations have led to identification of candidate genes for diseases including asthma, breast cancer, diabetes and cystic fibrosis. The emergence of capabilities to perform genome-wide association studies (GWAS) should also provide additional insight to the genetic basis of disease susceptibility. A weakness of many of these investigations is that generally one population is investigated. Replication is important because gene(s) may have varying importance between populations (i.e. disease allele frequencies may be different between populations, or may be too small to account for a major portion of disease incidence). It is also important to note that the influence of environmental exposures on disease phenotype(s) is not well understood in many instances and cannot be adequately evaluated or applied in many genetic studies; in chronic diseases this may be problematic.

A role for genetic susceptibility to the toxic effects of environmental stimuli has also been demonstrated. For example, animal modeling studies have identified quantitative trait loci (QTLs) and functionally relevant candidate genes for susceptibility to lipopolysaccharide (endotoxin) and ozone (O_3) (e.g. Kleeberger et al 1997, Poltorak et al 1998). Clinical and population studies in humans have also associated functionally relevant polymorphisms in candidate genes with adverse outcomes of exposure to these agents (e.g. Arbour et al 2000, Bergamaschi et al 2001). It is not clear whether genes that determine susceptibility to agents such as O_3 or particulates confer enhanced risk to development of diseases such as asthma, but it is interesting some genes such as *TNF* (tumor necrosis factor α) are important to pollutant responsiveness (e.g. Yang et al 2005) and chronic lung disease (e.g. Li et al 2006).

Individuals are also exposed to multiple environmental stimuli, either as complex mixtures or exposures, or as sequential events. While the importance of environmental exposures to disease is clear, they are difficult to quantitate. Tools are generally not sufficiently well-developed to determine precisely the acute, repeated or chronic exposures to environmental stimuli that may contribute to or influence disease pathogenesis. Furthermore, although we are exposed to environmental agents throughout our lifetime, the timing of exposures may impact disease. In particular, windows of vulnerability include *in utero*, infant and childhood exposures. Recent investigations of epigenetic effects on disease pathogenesis underscore the importance of understanding lifetime exposures to external stimuli (e.g. Jirtle & Skinner 2007).

Clearly, understanding each of these important contributing components is critical to treatment and prevention of environmental lung diseases. Unfortunately, gene × environment investigations are extremely costly, and labor and time intensive. Therefore, a clear need exists for predictive models of disease to help guide and focus genetic studies in human populations.

Primary cell culture systems and cell lines have proven extremely useful for manipulation of genes candidates that contribute to basal and environmental response phenotypes. The inbred mouse is also a particularly informative predictive model because genetic background is well controlled and the strong homology between mouse and human genomes. Traditional meiotic mapping techniques using strains of mice with disparate phenotypes for the same disease endpoint have identified QTLs that explain the differential phenotype of the parental strains (for review, see Cho & Kleeberger 2007). Recently developed emergent haplotype mapping algorithms based on high density SNP mapping across multiple inbred strains of mice have provided additional tools for investigators to identify disease genes (see e.g. Liao et al 2004, Pletcher et al 2004). The developing collaborative cross which will create 1000 recombinant inbred strains derived from eight parental strains should also greatly advance our ability to determine the genetic basis of disease phenotypes (Complex Trait Consortium 2004).

Identification of QTLs or chromosomal regions that explain a portion of the genetic variance in disease phenotypes is extremely informative, but often these chromosomal regions are relatively large and may contain hundreds of genes. Additional experimental approaches must be employed to determine the identity of a candidate gene or genes within the QTLs. Gene expression and proteomic technologies have enabled investigators to query disease QTLs and identify differential RNA and protein profiles that provide insight to the candidate gene or genes (e.g. Min-Oo et al 2007). Moreover, genetic manipulation of inbred strains of mice is feasible, including targeted disruption (tissue specific/global), overexpression, siRNA for silencing, and antibodies to gene products, to screen potential candidates. Mice with chromosome substitution ('congenic' or 'consomic' lines) also have been used to reduce the size of the QTLs. Comparative genomics across multiple species, including yeast, *Caenorhabditis elegans*, mouse, and humans, may be employed to identify conserved genes that may have relevance in disease pathogenesis (Kleeberger & Schwartz 2005). Ultimately, the long range goal for developing the predictive models is to understand human disease, and develop intervention strategies to prevent disease. The following describes a 'bench to bedside' investigation that initially used animal modeling to identify a gene for susceptibility to lung injury induced by oxidant exposure. Subsequent studies in cell lines characterized functional polymorphisms in the human homologue of the gene that were then found to be important in protection against oxidant injury in humans. This cross-species investigation illustrates the value of integrating multiple models to understand the interaction of genetic background and environmental exposure (i.e. gene × environment interaction) in disease causality.

Oxidant-induced lung injury

Reactive oxygen species (ROS) and oxidative stress have been increasingly implicated in the pathogenesis of many diseases and important biological processes including carcinogenesis, atherosclerosis, aging, neurodegenerative diseases, and inflammatory disorders (e.g. Ames & Shigenaga 1992, Floyd 1990, Jenner 1994). In addition, many pulmonary diseases (e.g. adult respiratory distress syndrome [ARDS], chronic obstructive pulmonary disease [COPD], bronchopulmonary dysplasia [BPD]) require supplemental oxygen therapy (hyperoxia) to maintain lung function which further increases the oxidant burden of the lung. It is believed that the damaging effects of oxygen are mediated by ROS including superoxide radical, H_2O_2, and hydroxyl radicals formed by membrane oxidase (e.g. NADPH oxidase) or mitochondria in epithelial cells, polymorphonuclear leukocytes (PMN), and macrophages, but the mechanisms are still not clear. Furthermore, identification of those factors that may influence susceptibility remains an important issue.

Because of the potential impact that oxidants may have on lung function, understanding of the susceptibility factors could lead to better intervention strategies and, potentially, a means to protect individuals at risk for the development of oxidative stress injury. Possible extrinsic factors include respiratory infection, nutritional status (especially antioxidant status), and previous inhalation exposure to toxic substances. Intrinsic or 'host' factors that may influence susceptibility to oxidant lung injury include the age of the individual, pre-existing disease state, and genetic background. Evidence that supports the role of genetic predisposition to hyperoxic lung injury can be inferred from the experience with BPD in premature infants (Clark & Clark 2005). Respiratory distress syndrome occurs secondary to surfactant deficiency and lung immaturity and is common in premature infants (Kinsella et al 2006). Treatment includes mechanical ventilation and oxygen, and many infants go on to develop a chronic form of lung disease, BPD (Kinsella et al 2006). Although a variety of factors have been implicated in the development of BPD, pulmonary barotrauma and exposure to high concentrations of inspired oxygen remain the most accepted causative factors (Kinsella et al 2006). Several studies have demonstrated a decreased incidence of BPD in black infants when compared with white infants matched for gestational age and birth weight. Similarly, males have a higher incidence of BPD than do gestational age- and sex-matched female infants (Avery et al 1987, Clark & Clark 2005). Although these findings may represent maturational differences, they also suggest that genetic background may be an important determinant in the susceptibility to hyperoxic lung injury and the process of recovery and repair. Further evidence implicating genetic susceptibility comes from Clark et al (1982) who demonstrated increased incidence of BPD among infants with the HLA-A2 haplotype. Premature infants with a family medical history for asthma in a parent or sibling are also more likely to develop BPD (Nickerson & Taussig 1980).

Results and discussion

We began to investigate the genetic basis of susceptibility to hyperoxic lung injury by determining the inter- and intra-strain variation in lung response to continuous exposure to hyperoxia (100% oxygen) among inbred strains of adult mice. Hyperoxic lung injury induces inflammation and noncardiogenic edema in the lung which are phenotypes of acute lung injury (ALI) and ARDS. Estimates of hyperoxia-induced alteration in lung permeability and inflammation were made by measuring the total protein content and counting the PMNs in bronchoalveolar lavage fluid (BALF). The interstrain variation in time course and magnitude of change in total protein concentration and PMNs was significantly greater than intrastrain variation among the strains of mice in response to hyperoxia (Hudak et al 1993). Susceptible strains (e.g. C57BL/6J, B6) developed 8–10-fold greater mean total protein in BAL returns compared to resistant strains (e.g. C3H/HeJ, C3). A mortality study indicated highly significant between-strain differences in the rate of lethality with continuous hyperoxia exposure (Fig. 1). For example, while no B6 mice survived beyond 80 h of hyperoxia, all resistant C3 and C57L/J mice survived beyond 96 h, and 50% survived over 120 h. It is interesting that C3 and C57L/J mice survived despite significant changes in lung permeability observed at this time. However, no strains were completely tolerant to hyperoxia, as all mice eventually succumbed to the exposure.

A genome-wide linkage analysis was then done with BXH Recombinant Inbred (RI) strains derived from B6 and C3 mice. Interval analyses identified suggestive QTLs on chromosomes 2 and 3, and independent confirmation of the QTLs was found with a cohort of B6C3F$_2$ mice that were phenotyped for reponsiveness to hyperoxia (Cho et al 2002a). A susceptibility locus for the PMN phenotype was identified on chromosome 2 (Cho et al 2002a). This major QTL overlapped the suggestive QTL on chromosome 2 for the BXH RIs. An additional, suggestive QTL was detected on chromosome 3. We designated the chromosome 2 and 3 QTLs as hyperoxia susceptibility locus 1 (*Hsl1*) and 2 (*Hsl2*), respectively. We found no significant interaction between QTLs for any phenotype, suggesting that the two QTLs independently affect the hyperoxia response.

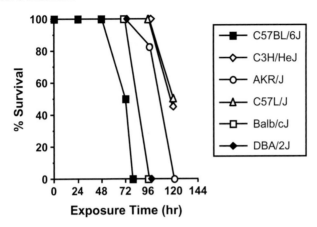

FIG. 1. Time course of hyperoxia-induced mortality in six inbred strains of mice. Symbols represent percent survival for each strain. Sample sizes at the beginning of exposure ranged from 18 (DBA/2J) to 36 (C57BL/6J).

Subsequent fine-mapping of *Hsl1* identified the greatest linkage between approximately 45 and 47 cM. A potential candidate gene within this QTL is nuclear transcription factor *Nrf2* (NF-E2 related factor 2; also called *Nfe2l2*, nuclear factor erythroid 2, like 2). NRF2 is an essential regulator of antioxidant enzymes and defense proteins for host protection against carcinogenicity, mutagenicity and oxidative stress (Itoh et al 1997, Chan et al 2001, Ramos-Gomez et al 2001). To test the hypothesis that *Nrf2* confers differential susceptibility to oxygen toxicity, NRF2-DNA binding activity (electrophoresis mobility shift assay [EMSA]) was evaluated in the lungs of C3 and B6 mice exposed to hyperoxia. NRF2 activity was markedly decreased in B6 mice by 6 h of hyperoxia, and continued to decrease during the exposure (Cho et al 2002a). In contrast, after 90 min exposure NRF2 activity increased in C3 mice proportional to gene expression and remained elevated until 48 h (Cho et al 2002a). To further investigate the role of *Nrf2* in differential susceptibility to hyperoxic lung injury, we have begun a sequence analysis of the gene in B6 and C3 mice. We identified a T→C substitution at the −336 site of the B6 *Nrf2* promoter that is predicted to have a functional effect on regulation of *Nrf2* expression (Cho et al 2002a). This single nucleotide polymorphism (SNP) cosegregated with susceptibility in the F_2 cohort, such that B6C3F$_2$ mice which were wild-type or heterozygous for the polymorphism were resistant to the effects of hyperoxia exposure, while those mice that were homozygous for the polymorphism were susceptible (Fig. 2). This segregation analysis was therefore consistent with a role for the SNP in susceptibility to hyperoxia-induced lung injury, and validated *Nrf2* as a susceptibility gene in pulmonary pathogenesis by hyperoxia. Further, relative to *Nrf2*$^{+/+}$ controls, mice with site-directed mutation (knock-out)

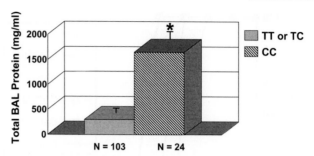

FIG. 2.   Effect of a T→C substitution at the −336 site of the *Nrf2* promoter on hyperoxia-induced change in bronchoalveolar lavage (BAL) protein obtained from a cohort of B6C3F$_2$ mice. Mean BAL protein levels were significantly greater in mice that were CC (homozygous mutant) for the polymorphism compared to TT (wild-type) or TC (heterozygous) mice. Mean S.E.M are presented. *, $P < 0.05$.

of *Nrf2* (ICR-*Nrf2*$^{-/-}$) were significantly more susceptible to hyperoxic lung injury (Cho et al 2002b). These mice were also more sensitive to bleomycin-induced fibrosis (Cho et al 2004), and cigarette-smoke-induced lung injury (Rangasamy et al 2004), suggesting a significant role for NRF2 in pulmonary responses to oxidant stressors in adult mice. Microarray analysis of *Nrf2*- and hyperoxia-dependent gene expression in whole lung homogenates demonstrated potential antioxidant defense pathways through which *Nrf2* confers protection against oxidant lung injury (Cho et al 2005).

Based on the findings from mouse studies in which the candidate susceptibility gene *Nrf2* is essential to protection against oxidative pulmonary injury (e.g. Aoki et al 2001, Cho et al 2002b, Cho et al 2004, Rangasamy et al 2004), we hypothesized that polymorphisms in *NRF2* resulting in decreased function similarly predispose humans to ALI. To address this hypothesis, we first resequenced *NRF2* in four different ethnic populations and identified three new *NRF2* promoter polymorphisms at positions −617 (C/A), −651 (G/A) and −653 (A/G) (Marzec et al 2007). Transcription factor motif analysis (TRANSFAC) indicated that variations at −653/−651 and −617 alter the consensus recognition sequences for MZF1 (Myeloid zinc finger-1) and NRF2, respectively, suggesting that these polymorphisms may affect *NRF2* transcription. To examine this concept, we transiently transfected A549 cells with the vector constructs 727-Luc (inclusion of the SNPs) or 538-Luc (deletion of the SNPs) and basal level activity of the *NRF2* promoter with or without polymorphisms was determined. Compared to the basic pGL3-vector, the 538-Luc displayed significantly higher level of activity ($P < 0.001$). The −538 to +131 bp region of *NRF2* (538-Luc) contains regulatory elements that drive the promoter activity above the levels seen with the promoter-less basic vector. However, the inclusion of −538 to −727 region to the 538-Luc construct

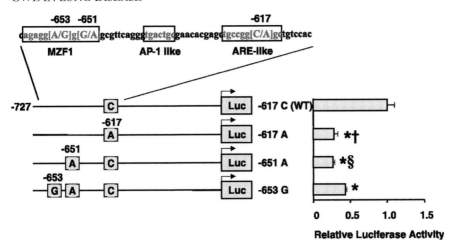

FIG. 3.   Effects of *NRF2* polymorphisms on *NRF2* promoter activity. Luciferase activity of four *NRF2* promoter constructs: −617C (wild-type sequence, 727-Luc), −617 A, −651 A, and −653 G. Remaining SNP sites are indicated when double variants are used. Wild-type 727-Luc reporter exhibited >50% higher luciferase activity compared to polymorphic variants. * *P* < 0.001 compared to −617 C (WT); †*P* = 0.014 compared to −651 A; §*P* = 0.019 compared to −653 G. *P* values were calculated from a Student-Newman-Keuls *a posteriori* comparisons of means test after one-way analysis of variance; group sizes = 4. (Adapted from Marzec et al 2007.)

(i.e. 727-Luc) further enhanced the reporter activity (Marzec et al 2007). These results indicate that the −538 to +131 bp region acts as a basal promoter, while the −538 to −727 region acts as an enhancer. Relative to 538-Luc, the 727-Luc reporter had fourfold higher luciferase activity (*P* < 0.001) indicating that the −538 to −727 region most likely contains DNA sequences required for high level *NRF2* promoter activity (Fig. 3). Finally, we tested the association of functional SNPs (−617, −651) with differential risk for ALI in patients following major trauma. Patients with the −617 A SNP had a significantly higher risk for developing ALI after major trauma (OR 6.44; 95% CI 1.34, 30.8; *P* = 0.021) relative to patients with the wild type (−617 CC) (Marzec et al 2007). This translational investigation provides novel insight into the molecular mechanisms of susceptibility to ALI, and may help to identify patients who are predisposed to develop ALI under at risk conditions, such as trauma and sepsis.

Summary

It has become clear that interaction between genetic background and exposures to environmental stimuli underlie the etiology of environmental-associated lung disorders. Although the mechanisms of multiple environmental stimuli on genetic background during the disease pathogenesis remain to be understood, application of genetic linkage analysis in mice and disease cohort studies in humans have provided insight to these interactions. We described a murine model of oxidative pulmonary injury to investigate genetic mechanisms of ALI/ARDS susceptibility. Genome-wide scans in inbred mice have identified chromosomal QTLs that include candidate susceptibility genes that confer differential susceptibility to oxygen toxicity in this model. Functional studies with gene knockout mice, sequence analysis, and global gene expression analysis have strongly supported the role of a novel candidate gene. Because considerable homology exists between the human and mouse genomes in the significant QTL, the loci (or the genes) in regions may have relevance in human responses. Subsequently, the candidate susceptibility gene identified from the murine model was investigated in human disease cohorts, and the results have added knowledge of the pathogenesis in ARDS/ALI as well as other oxidative lung disorders. We believe that understanding of gene–environment interactions in complex lung diseases by 'bench to bedside' investigations will be essential to the development of new strategies for prevention and treatment of environmentally relevant pulmonary disorders.

Acknowledgements

This work was supported by the Intramural Research Program at the National Institute of Environmental Health Science, National Institutes of Health.

References

Ames BN, Shigenaga MK 1992 Oxidants are a major contributor to aging. Ann N Y Acad Sci 663:85–96

Aoki Y, Sato H, Nishimura N, Takahashi S, Itoh K, Yamamoto M 2001 Accelerated DNA adduct formation in the lung of the Nrf2 knockout mouse exposed to diesel exhaust. Toxicol Appl Pharmacol 173:154–160

Arbour NC, Lorenz E, Schutte BC et al 2000 TLR4 mutations are associated with endotoxin hyporesponsiveness in humans. Nat Genet 25:187–191

Avery ME, Tooley WH, Keller JB et al 1987 Is chronic lung disease in low birth weight infants preventable? A survey of eight centers. Pediatrics 79:26–30

Bergamaschi E, De Palma G, Mozzoni P et al 2001 Polymorphism of quinone-metabolizing enzymes and susceptibility to ozone-induced acute effects. Am J Respir Crit Care Med 163:1426–1431

Chan K, Han XD, Kan YW 2001 An important function of Nrf2 in combating oxidative stress: detoxification of acetaminophen. Proc Natl Acad Sci USA 98:4611–4666

Cho HY, Kleeberger SR 2007 Genetic mechanisms of susceptibility to oxidative lung injury in mice. Free Radic Biol Med 42:433–445

Cho H, Jedlicka AE, Reddy SP, Zhang LY, Kensler TW, Kleeberger SR 2002a Linkage analysis of susceptibility to hyperoxia: Nrf2 is a candidate gene. Am J Respir Cell Mol Biol 26:42–51

Cho HY, Jedlicka AE, Reddy SPM et al 2002b Role of NRF2 in protection against hyperoxic lung injury in mice. Am J Respir Cell Mol Biol 26:175–182

Cho HY, Reddy SPM, Yamamoto M, Kleeberger SR 2004 The transcription factor NRF2 protects against pulmonary fibrosis. FASEB J 18:1258–1260

Cho HY, Reddy SP, DeBiase A, Yamamoto M, Kleeberger SR 2005 Gene expression profiling of NRF2-mediated protection against oxidative injury. Free Radic Biol Med 38:325–343

Clark H, Clark LS 2005 The genetics of neonatal respiratory disease. Semin Fetal Neonat Med 10:271–282

Clark DA, Pincus LG, Oliphant M, Hubbell C, Oates RP, Davey FR 1982 HLA-A2 and chronic lung disease in neonates. JAMA 248:1868–1869

Complex Trait Consortium 2004 The collaborative cross, a community resource for the genetic analysis of complex traits. Nat Genet 36:1133–1137

Elias JA, Lee CG, Zheng T, Ma B, Homer RJ, Zhu Z 2003 New insights into pathogenesis of asthma. J Clin Invest 111:291–297

Floyd RA 1990 Role of oxygen free radicals in carcinogenesis and brain ischemia. FASEB J 4:2587–2597

Hudak BB, Zhang L-Y, Kleeberger SR 1993 Inter-strain variation in susceptibility to hyperoxic injury of the murine lung. Pharmacogenetics 3:135–143

Itoh K, Chiba T, Takahashi S et al 1997 An Nrf2/small Maf heterodimer mediates the induction of phase II detoxifying enzyme genes through antioxidant response elements. Biochem Biophys Res Commun 236:313–232

Jenner P 1994 Oxidative damage in neurodegenerative disease. Lancet 344:796–798

Jirtle RL, Skinner MK 2007 Environmental epigenomics and disease susceptibility. Nat Rev Genet 8:253–262

Kinsella JP, Greenough A, Abman SH 2006 Bronchopulmonary dysplasia. Lancet 367:1421–1431

Kleeberger SR, Peden D 2005 Gene–environment interactions in asthma and other respiratory diseases. Annu Rev Med 56:383–400

Kleeberger SR, Schwartz DA 2005 From QTL to gene: a work in progress (editorial). Am J Respir Crit Care Med 171:804–805

Kleeberger SR, Levitt RC, Zhang L-Y et al 1997 Linkage analysis of susceptibility to ozone-induced lung inflammation in inbred mice. Nat Genet 17:475–478

Li YF, Gauderman WJ, Avol E, Dubeau L, Gilliland FD 2006 Associations of tumor necrosis factor G-308A with childhood asthma and wheezing. Am J Respir Crit Care Med 173:970–976

Liao G, Wang J, Guo J, et al 2004 In silico genetics: identification of a functional element regulating H2-Ealpha gene expression. Science 306:690–695

Marzec J, Christie JD, Reddy SP et al 2007 Functional polymorphisms in the transcription factor NRF2 in humans increase the risk of acute lung injury. FASEB J 21:2237–2246

Min-Oo G, Fortin A, Pitari G, Tam M, Stevenson MM, Gros P 2007 Complex genetic control of susceptibility to malaria: positional cloning of the Char9 locus. J Exp Med 204:511–524

Nickerson BG, Taussig LM 1980 Family history of asthma in infants with bronchopulmonary dysplasia. Pediatrics 65:1140–1144

Pletcher MT, McClurg P, Batalov S et al 2004 Use of a dense single nucleotide polymorphism map for in silico mapping in the mouse. PLoS Biol 2:2159–2169

Poltorak A, He X, Smirnova I et al 1998 Defective LPS signaling in C3H/HeJ and C57BL/10ScCr mice: mutations in Tlr4 gene. Science 282:2085–2088

Ramos-Gomez M, Kwak MK, Dolan PM et al 2001 Sensitivity to carcinogenesis is increased and chemoprotective efficacy of enzyme inducers is lost in nrf2 transcription factor-deficient mice. Proc Natl Acad Sci USA 98:3410–3415

Rangasamy T, Cho CY, Thimmulappa RK et al 2004 Genetic ablation of Nrf2 enhances susceptibility to cigarette smoke-induced emphysema in mice. J Clin Invest 114:1248–1259
Yang IA, Holz O, Jorres RA et al 2005 Association of tumor necrosis factor-alpha polymorphisms and ozone-induced change in lung function. Am J Respir Crit Care Med 171:171–176

DISCUSSION

Martin: Would it be worth looking at these genes in other respiratory diseases such as COPD or even asthma?

Kleeberger: Yes. We are currently looking at emphysema. A recent paper demonstrated in a model of cigarette smoke exposure in mouse that NRF2 is important in emphesymatous lesions that occurred (Rangasamy et al 2004). My lab is also evaluating the role of NRF2 polymorphisms in an emphysema cohort in Korea. So far it is looking promising. COPD would be another important group to work with. We are also evaluating the role of NRF2 in a sepsis cohort, and we are developing a cohort of bronchopulmonary dysplasia (BPD) kids. The model of hyperoxic lung injury that I just described was started by Dr Bonnie Hudak, who was a pediatric fellow at Hopkins with an interest in BPD.

Martin: Why not just look at lung cancer?

Kleeberger: People are looking, including ourselves. It is emerging that NRF2 may be considered a double-edged sword. It contributes to carcinogenesis in the lung. *Nrf2* knockout mice are highly protected against the tumorigenesis in the lung induced by chemicals. We are not clear why this is, but it is being reproduced in other laboratories. It may be that NRF2 is involved in cell cycle regulation, and it is in some way contributing to pathogenesis rather than protecting against it. The jury is still out in terms of in terms of all the things that NRF2 can do in protection against lung injury and cancer. Having said this, there are many studies that have investigated the role of NRF2 in liver cancer induced by alfatoxin. In rat and mouse models induction of NRF2 with oltipraz protects against liver cancer induced by alfatoxin. This may be a timing effect or an organ-specific effect.

Braithwaite: Oxidative stress induces p53, and in humans p53 mutations are common in lung cancers. Have you crossed your *Nrf2* mice onto a p53 heterozygote background to see whether this changes the frequency?

Kleeberger: No. It should be done, though. We have begun to take the same strategy with an innate immune gene, *Tlr4*. In our lab we found there is some interaction between NRF2 and TLR4 in ways that can modulate inflammation in injury in the lung. We have begun back-crossing the *Nrf2* knockout onto a mutant background for *Tlr4*, as well as the wild-type background, to help us evaluate the interaction of these two genes in oxidant injury of the lung. The role of p53/p51 will also be important to investigate.

Braithwaite: Is apoptosis important in an inflammatory response? Is it protective or does it enhance disease?

Kleeberger: Acute lung injury is mostly a necrotic effect. If we create a different model with hyperoxia, we get more of an apoptotic phenotype, as well as the necrosis. We haven't pursued the role of NRF2 in that model yet.

Rutter: I'd like to make a different sort of comment. You used the term 'translational research'. One of the issues in the UK at the moment is that some policy people seem to think that laboratory science provides all the ideas, and that translation is restricted to randomized controlled trials to get the treatment to the bedside. Harold Himsworth (1962), who was head of the MRC at the time, argued strongly that the distinction between basic and applied research is misleading. The two are on a continuum and the kind of thinking involved in the two circumstances is very similar. Keith Peters (2004) made a somewhat different point: he argued that the notion that this is one-way traffic is completely wrong. Many of the ideas start from the clinical science, go to basic science and then come back again. It is a kind of iterative process. A better terminology is to talk about this as 'experimental medicine'. This is at risk, at least in the UK. You start, as did Malak Kotb, with an observation of marked individual human variation in particular kinds of outcomes. This leads you to think what might be the biological processes. Then you used a range of investigations using different strategies, some of which are better described as basic science and some of which are not. This is a false dichotomy.

Kleeberger: I completely agree with your comments. 'Translational' is terribly vague. We have just initiated a program that is based on multiple ways of investigating the same question, using multiple species (mouse, human, *C. elegans*, yeast), bioinformatics and computational biology, all of which feed into each other in terms of generating hypotheses that can be pursued in all groups. Is there one road? No, it can't be a single road. It is a feed-forward, feed-backward process that we need to embrace. The point I was trying to make here is that one direction (from mouse to human) can be an informative strategy; we don't always think about the mouse as being a good predictive model of human disease.

Kotb: You have made an important point. I have the good fortune of working closely with clinical people. There is no way we could have thought or designed our studies had it not been for their input. I have noticed how naïve some of my designs would have been if it had not been for the clinical input. It is definitely a back and forth process. When both sides are sharing ideas, and there is no ego, it is a beautiful process. In terms of terminology, we are probably stuck with the term 'translational'. There is research that has application because it is trying to find a mechanism, and there is research which is also trying to find the mechanism but with a very specific purpose. For example, we want to do individualized medicine. We want to make something that the clinical person will take and use. In addition, to get there we want to understand the mechanism of disease, but our focus keeps on being how we are going to use this information in patients.

Reeve: In NZ the health research council has been debating what the term translational research means for the last 18 months. As a biomedical scientist who was there when the word was invented, for me it means from bench to bedside. We have debated this in a wider group. It was a complex definition in the end because everyone had their oar in the process. One was that translational research has to have the intention of uptake. I think this is important. Uptake to whom? To the clinic and the community. As medical researchers we think it will just be the doctor in the office next door who will use the test, but it actually had a wider connotation than this. There also had to be a timeline on this, which I didn't entirely agree with. I think translational research could occur over a period of 15 years. The definition specified that this research had to be completed within the short to medium term.

Rutter: You are right about the time frame. It is naïve to suppose that all of this can happen in the short term. I presume from the way you put it, that the term 'useful' implies public health implications, and not just treatment of individual patients.

Reeve: Yes.

References

Himsworth H 1962 Society and the advancement of natural knowledge. Br Med J 2:1557–1563
Peters K 2004 Exceptional matters: clinical research from bedside to bench. Clin Med 4:551–566
Rangasamy T, Cho CY, Thimmulappa RK et al 2004 Genetic ablation of Nrf2 enhances susceptibility to cigarette smoke-induced emphysema in mice. J Clin Invest 114:1248–1259

GENERAL DISCUSSION III

Rutter: Richie Poulton, I'd like to return to your presentation. As I understood the messages you were putting forward, you were dealing with two rather separate challenges. The first was the accusation that the focus on genetics was getting in the way of environmental interventions that will solve the problem. The answers you gave on this were persuasive, but I wonder whether they go far enough? For example, what simply looking at making environments better doesn't tell you is the developmental moderation of many risks. It does matter <u>when</u> things happen as well as whether they happen. Equally, there is an assumption that removing the risks is what it is all about. But the analogy with vaccination makes one pause here. Challenges (whether infectious or psychosocial) are part of normal function. If you want to protect people against infectious disease, you don't prevent them from having any exposure. The example of Tristan de Cunha (an isolated island in the Pacific) showed the problems of this. The islanders fled when there was a volcanic eruption, came to England and suffered heavily from infectious disease because they hadn't had any previous exposure. The issue is not simply one of removal or risk; it is a question of exposure to risk in ways that the organism can take account of. Gene–environment interplay is one way of looking at that. The third issue that I would raise is that the assumption that you can intervene in risk without knowing the origins of the risk, seems to be starting in the middle. One needs to ask not only whether there is a risk, but why some individuals have a horrendous series of risks to which they are exposed to, and others don't. This takes us into gene–environment correlations rather than interactions, but it is part of the same thing. The fourth point I would make is that made by Ceci and Papierno (2005). They asked that if there is universal provision of beneficial environments, this may make inequalities worse, because of differential take-up of the opportunities. How do we deal with this? One way of dealing with this is to make sure that everyone has a bad environment so that the issue doesn't arise! If we put that aside, one needs to know something about the mechanisms involved in risk to see how we can counter this in a way that is effective. Simply blasting in with improving the environment, desirable though that is, is not enough.

Then you were taking on a different challenge; the question of why we need longitudinal studies. There were two things I would comment on here. First, the big advantage of a longitudinal study is the ability to look at unexpected outcomes. Case control studies cannot do that. We have to define what a case is at the start. The sort of research I have done over many years has thrown up many unexpected outcomes. The notion that institutional deprivation would lead to autistic-like patterns would not have occurred to me in a month of Sundays, but it does (Rutter et al 2007a). Whether this is dealing with 'ordinary' autism is a moot point, but it is saying that you have a highly specific outcome of an unexpected kind. In line with this, you talked about longitudinal studies as if they needed to be general population. I would argue that there are many reasons why you might want a full unselected population also you may need to investigate rare outcomes. These may be high risk defined in family history as in the Edinburgh schizophrenia study (Owens & Johnstone 2006) looking at the translation from genetic liability to overt schizophrenia, or it may be a serious environmental risk such as our study of children in Romanian institutions (Rutter et al 2007b). The arguments for using genetics to understand environments (see Rutter 2007) are something that the public needs to be educated on.

Martinez: I would like to pose a problem that I was faced with. I was a member of an advisory board for the national heart, lung and blood institute in the USA. We reviewed projects that were presented to us by staff of the institute for funding. Two years ago a couple of projects were proposed to us that were going to cost US$50 million in total. The proposal was basically to determine whether it was cost effective to treat people with subclinical levels of cholesterol in the circulation, and subclinical hypertension. Since there is a small risk of bad outcomes for these people, it was proposed that we treat a sample of subjects with, say, a diastolic pressure of 85 mm Hg. The idea was that perhaps what we have to do to decrease mortality further is to start moving down the hypertension–mortality curve. Many of us said, what is the limit? You would end up treating 40 or 50 people to avoid one death. The proposal was thus made to try to identify within the group of individuals who have these borderline characteristics who are the ones at the highest risk of dying. For example, the relative risk for dying in the group of patients with a diastolic pressure of 85 is perhaps 1.2, whereas with 110 it is probably 3. Perhaps we need to move towards something that you are hinting at: we all know that it is best not to be mistreated when you are a child, but we don't know really whether everyone has to be thin, which is what we are telling everyone to be as a society. I mentioned earlier the paper on overweight status (Flegal et al 2007) showing that globally the people with a BMI between 25 and 30 have less mortality than the rest of the population. They have more cardiovascular disease but less cancer. Perhaps we could identify the people who would do better from the point of view of cancer, but not from the point of view of cardiovascular disease. This should be the focus of our attention: these kinds of intermediate levels of risk, where we are not sure what we should do in terms of environmental exposures, such as dieting or exercising extensively. We would all like to be better off, but maybe we don't all need to diet!

Rutter: It is a great example. There has been a lot of discussion on this, which starts with the evidence that cholesterol levels are a risk factor right across the distribution and therefore everyone should be treated. Then we get to the question of risks versus benefits (see Academy of Medical Sciences 2007). Statins are pretty safe drugs but they are not totally free of side-effects. Using them on individuals who have no substantial risk it is a different matter to using them on people with high risk. This is where a study of gene–environment interactions could tell you what to do. The problem that we are in the middle of now is that we can all see the potential, but it is not quite ready for delivery in the way that some politicians would like.

Braithwaite: You could get much the same information from family history.

References

Academy of Medical Sciences 2007 Identifying the environmental causes of disease: how should we decide what to believe and when to take action? Academy of Medical Sciences, London

Ceci SJ, Papierno PB 2005 The rhetoric and reality of gap closing: when the 'have-nots' gain but the 'haves' gain even more. Am Psychol 60:149–160

Flegal KM, Graubard BI, Williamson DF, Gail MH 2007 Cause-specific excess deaths associated with underweight, overweight, and obesity. JAMA 298:2028–2037

Owens DGC, Johnstone EC 2006 Precursors and prodromata of schizophrenia: findings from the Edinburgh High Risk Study and their literature context. Psychol Med 36:1501–1514

Rutter M 2007 Proceeding from observed correlation to causal inference: the use of natural experiments. Perspect Psychol Sci 2:377–395

Rutter M, Kreppner J, Croft C et al 2007a Early adolescent outcomes of institutionally deprived and non-deprived adoptees. III Quasi-autism. J Child Psychol Psychiatry 48:1200–1207

Rutter M, Beckett C, Castle J, Colvert E, Kreppner J, Mehta M 2007b Effects of profound early institutional deprivation: an overview of findings from a UK longitudinal study of Romanian adoptees. Eur J Dev Psychol 4:332–350

12. Gene–environment interaction in complex diseases: asthma as an illustrative case

Fernando D. Martinez

Arizona Respiratory Center, University of Arizona, 1501 N Cambpbell Ave., Rm 2349, PO Box 245030, Tucson, AZ 85724-5030, USA

Abstract. Genetic studies of asthma have been plagued by a remarkable difficulty in constantly replicating results in different populations for most of the polymorphisms studied. This was true even when the quality of the study design and statistical power were not an issue. The most plausible explanation for these inconsistent results is that genetic polymorphisms, in most cases, do not directly influence risk for asthma but instead modulate the effect of environmental exposures on the inception and clinical expression of asthma and allergies. A better understanding of the genetics of asthma is thus inseparable from a better understanding of the mechanisms by which environmental factors increase the risk for asthma or protect against it.

2008 Genetic effects on environmental vulnerability to disease. Wiley, Chichester (Novartis Foundation Symposium) p 184–197

The development of new technologies to screen hundreds of thousands of genetic polymorphisms spanning the whole genome has engendered new hope that soon a map of genetic variations associated with different complex diseases such as allergies and asthma will become available. The premise behind this renewed hope is that there is a non-ambiguous, at least partially linear relationship between alleles in genetic variants in the genome and the risk of developing these diseases. In other words, geneticists believe that there is a set of genetic variants in the human genome that predispose for asthma and allergies, and that specific alleles for these variants will have an effect on asthma prevalence regardless of other factors such as environmental exposures or the influence of other variants.

This premise has not been entirely supported by efforts conducted in many locales using family-based approaches (linkage analyses) and candidate gene approaches. With these two types of study design, researchers initially described a large number of chromosomal regions (in the case of linkage studies) and dozens of single nucleotide polymorphisms that, in specific populations, were either linked or associated with asthma and allergies (reviewed in Los et al 1999, Ober & Hoffjan 2006). However, when corroboration of these findings was attempted in other populations, often with very large samples, results were either mixed or disappointing: in the case of asthma and allergies for example, less than 10% of all reported association 'hits' were replicated consistently, and even in these cases, replication was not observed in several well designed, sufficiently large samples.

Why the difficulty to replicate?

Two main types of interpretations have been offered for this consistent inconsistency. The most common ones, especially among quantitative geneticists, have been technical; perhaps the original observations were spurious, simply because scientists published only results for those comparisons that were 'positive', whereas negative results remained unreported. Therefore, the original reports suffered from type I error, and this is revealed by the replicating studies, in which no association is observed. A second possible technical explanation is that the replicating studies were not large enough and therefore, suffered of a type II type of error. Finally, it is often the case that the different studies are often not designed with the specific objective of corroborating each others findings but for other purposes. It is thus possible (even likely) that the phenotypes of interest were not equally defined in the original studies and in the replicating studies. A good example of how this may affect results of association studies is offered by the parallel study of the association between asthma and vitamin D receptor (VDR) polymorphisms (Raby et al 2004, Poon et al 2004). Raby et al (2004) and Poon et al (2004) reported simultaneously that the single nucleotide polymorphism (SNP) rs7975232 in VDR was associated with asthma, but the same allele (C) that was associated with increased risk of asthma in one of the populations in the Raby study and in the only population assessed in the Poon study was protective against asthma in the second population assessed in the Raby study. The authors (Raby et al 2004) proposed that perhaps the asthma phenotype studied in the children involved in the study where the C allele was protective was different from that of the mostly adult populations included in the studies showing that the C allele for rs7975232 increase the risk of asthma. In any event, the way in which an allele that causes increased risk of asthma in one population can protect against asthma in a second one remained unexplained. In addition, this set of studies hinted to an even more complex problem: had the authors studied all subjects together, regardless of age and locale, the effects in one direction in one population would have masked those on the other direction in the other population and they would have had to conclude that VDR was not a risk factor for asthma.

The above considerations notwithstanding, these types of technical explanations do not question the essential premise exposed above, that is, that there are indeed variants in the genome for which one specific allele will reproducibly show higher risk for complex diseases than the alternative allele. With this premise, the final objective of genetic studies of complex diseases is to find genetic variants that, when tested in random subjects, would inform us if the subject is at increased risk of the disease regardless of the context in which the disease occurs, much as today is done for single gene diseases such as cystic fibrosis or Tay-Sachs.

I have proposed an alternative explanation for the consistent difficulty to reproduce the results of well-designed genetic studies of complex diseases (Martinez 2007a): the effect of most genetic variants on the risk of asthma, allergies and other complex diseases is most often context dependent, and this 'context' may be genetic (gene–gene interactions or epistasis), environmental (gene–environment interactions), ethnic (which comprises both the genetic and environmental factors that makeup what is called an ethnic group), or developmental (which has to do with the timing of the combined effects that makeup a new phenotype).

The gene–environment interaction paradigm

Elsewhere I have provided a detailed description of the basic premises of the gene–environment interaction paradigm (Martinez et al 2007b). My assumption is that in most complex diseases such as asthma, environmental factors play a crucial role. This assumption is not only supported by twin studies showing that at least 50% of the risk for asthma (used herein as an illustrative case) is explained by environmental exposures (Duffy et al 1990), but also by the simple clinical observation that the expression of the disease is heavily influenced by environmental factors such as respiratory infections, allergens, emotions, air pollution, cigarette smoke, etc. Asthma also runs in families, and there is strong evidence in favor of a hereditary component of the disease. The decisive issue then becomes what proportion of the genetic variance is explained by a direct effect of polymorphisms on the phenotype regardless of the known environmental influences, and what proportion of the same genetic variance exerts its influence by modulating the effects of the environmental exposures. The implicit premise of genetic studies that assess potential direct effects of polymorphisms on a phenotype regardless of exposures is that there are indeed many polymorphisms in the genome that directly determine risk for the phenotype of interest, and in the same direction in all populations. I have alternatively proposed that, at least on the case of asthma and allergies, what most genetic polymorphisms do is to make the individual more or less susceptible to the environmental factors that are associated with the inception or expression of the disease.

To support this contention, I have extensively reviewed the experience by our group and others regarding the association between a polymorphism in the CD14 gene (CD14/−0159) and the development of asthma and allergies (Martinez

2007c). This polymorphism was originally described by our group in 1999 (Baldini et al 1999), and at that time we observed less severity of allergy in homozygotes for the T-allele for CD14/–159 as compared with carriers of the other two genotypes in a population studied in Tucson, Arizona. As with many other association studies, this finding was replicated by several groups but was not replicated in a similar number of studies. Interestingly, Kedda et al found no association between CD14/–159 and asthma and allergies in a meta-analysis of most studies published until 2005, but did report significant between-study heterogeneity (Kedda et al 2005).

At the same time that these genetic association studies were being performed, researchers studying asthma in rural communities in Central Europe observed that children of farmers exposed to high levels of endotoxin in house dust were significantly less likely to have allergic sensitization and atopic asthma than children living in the same rural communities but with low exposure to endotoxin (Braun-Fahrlander et al 2002). These studies in rural Europe provided strong support, therefore, for the so-called hygiene hypothesis, because they suggested that exposure to bacterial products deviated immune responses away from those that determine the allergic phenotype. We reasoned that, because CD14 is a crucial component of the receptor system for endotoxin, CD14 could mediate the protective effect of endotoxin on allergic sensitization. Therefore, in collaboration with the groups in Central Europe that had initially reported the farmers' studies, we genotyped the same children enrolled in those studies for CD14/–159. We found striking differences in the protective effect of endotoxin against allergies in carriers of different genotypes for this polymorphism. Homozygotes for the C-allele showed a steep, negative relationship between exposure to endotoxin and risk for allergies, whereas very little protective effect of endotoxin was observed among homozygotes for the T-allele. This different susceptibility to environmental exposures in subjects with different genotypes for CD14/–159 resulted in a peculiar pattern of genetic association: at low levels of endotoxin exposure, it was CC homozygotes that were at higher risk for allergies than TT homozygotes; conversely, at high levels of exposure it was TT homozygotes that were at higher risk for allergic sensitization than CC homozygotes. This pattern of opposite affects of the same allele in different environments/populations is similar to that reported for the vitamin D receptor, as explained earlier. It is thus possible that the apparently contradictory results obtained in different studies for genetic association may be due to different patterns of gene–environment interaction in those different locales.

Our finding of a complex gene–environment interaction between CD14 and endotoxin exposure as determinants of allergic sensitization has been replicated in a large study of children enrolled in a newborn cohort study in Manchester, UK (Fig. 1) and in at least two other studies in Barbados, West Indies, and Detroit

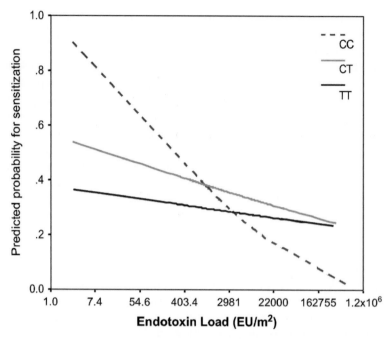

FIG. 1. Association between allergic sensitization at the age of 5 years and exposure to endotoxin in house dust by CD14/–159 genotype. It is apparent that the protective effect of endotoxin exposure is much stronger among CC homozygotes than among carriers of the other two genotypes. As a result, CC homozygotes are at risk for allergic sensitization at low levels of exposure and protected against sensitization at high levels of exposure as compared to the other two genotypes. At intermediate levels of exposure to endotoxin, there is no association between genotype and allergic sensitization.

in the USA (Zambelli-Weiner et al 2005, Simpson et al 2006, Williams et al 2006). This has prompted us to propose the hypothesis that, for most polymorphisms associated with asthma, the true role of the different genotypes for that polymorphism will only be completely understood if the corresponding set of environmental factors that interact with that particular polymorphism is also understood.

Results of recent genome-wide association studies

The recent flurry of studies using genome-wide association (GWA) approaches, which have been published in all major scientific journals and include one in reference to childhood asthma published in *Nature* (Moffatt et al 2007), seems to negate this conclusion. In all of these studies, hundreds of thousands of SNPs were genotyped in thousands of individuals with and without the phenotypes of interest, and at least one SNP was found to be associated with such phenotypes in each of these studies. Moreover, the polymorphism was replicated in anywhere between two and a dozen or more populations. At first sight, these results would thus suggest that there may be many SNPs that determine asthma risk independently of any environmental exposure.

The case of *ORMDL3* and its association with childhood asthma is a very cogent example of this new wave of GWA studies. Moffatt and coworkers (Moffatt et al 2007) characterized more than 317 000 SNPs in DNA from 994 patients with childhood onset asthma and 1243 non-asthmatics using family and case reference panels. They found multiple markers on chromosome 17Q10D1 that were strongly associated with childhood onset asthma, with a combined *P*-value of $P < 10^{-12}$. They replicated this finding in a cohort of 2320 German children ($P = 0.0003$) and in 3301 subjects from the British 1958 Birth Cohort ($P = 0.00005$). *ORMDL3* is the third member of a novel class of genes of unknown function that encode transmitter proteins anchored to the endoplasmic reticulum. Why such a gene would be associated with asthma is currently unknown.

Although the data presented by Moffatt and coworkers strongly suggests that *ORMDL3* may indeed be a gene containing polymorphisms that affect asthma risk, it is prudent at this point to wait for further corroboration of these findings in other population samples. This cautious approach is supported by the results of attempts to replicate results obtained with a gene not related to asthma and allergies, *INSIG2*, which was initially found to be associated with body mass index and obesity in a paper published in 2006 in the journal *Science* (Herbert et al 2006).

The case of *INSIG2* and obesity

Herbert and coworkers genotyped over 116 000 SNPs in 694 participants in the Framingham Heart Study (Herbert et al 2006). They searched for associations between over 86 000 of these SNPs and body mass index (BMI). They found a polymorphism (rs7566605) located in the region of the *INSIG2* gene that showed clear, genome-wide evidence of association with BMI in their Framingham population. They next attempted replication of the findings in relation to BMI and were able to replicate the association in four separate samples composed of individuals of Western European ancestry, African-Americans and children. The association was not replicated in one sample, the Nurses Health Study in the USA, but the authors attributed this lack of replication to the fact that there were fewer obese subjects in that population.

Subsequent to this publication, however, three groups of researchers from the USA, from France and from the UK, in studies that involved a large number of subjects, reported that they were unable to replicate the original finding by Herbert et al (Loos et al 2007, Dina et al 2007, Rosskopf et al 2007). In addition, two separate publications from England (Smith et al 2007) and from India (Kumar et al 2007) also showed no association of the same polymorphism in *INSIG2* and obesity in their samples. In an attempt to explain these discrepancies, some of the authors of the original publication, in collaboration with other researchers, published a comprehensive analysis of the association between rs7566605 and either BMI and a continuous variable or obesity status in 9 large cohorts from 8 populations across multiple ethnicities, for a total of almost 17 000 subjects (Lyon et al 2007). The results suggested that, much like for the great majority of other studies of association based on candidate genes, published before the advent of genome-wide searches, the association could be observed in some populations but not in others. When the authors combined all results in a single analysis, the association between BMI and rs7566605 was borderline significant among unrelated subjects ($P = 0.046$) and significant in family based studies ($P = 0.0004$), but the strength of the combined effect of rs7566605 on BMI was very modest.

Non-linear causation in complex diseases

The reader will forgive the extensive analysis of association studies with a phenotype that is not strictly relevant to the main issue at hand, namely the genetics of asthma and allergies. However, the experience with *INSIG2* and obesity is very relevant because it shows that, at least in the specific case at hand, GWA studies are subject to the same problems and apparently contradictory results that had been previously observed for association studies of candidate genes. Perhaps the best example in the field of allergies provided by the study of Maier et al in relation to circulating total IgE (Maier et al 2006). These authors attempted to replicate in a single, large population (4570 individuals who participated in the British 1958 Birth Cohort) the results previously reported in relation to allergies for SNPs in eight genes that had been previously reported to be associated with susceptibility for high IgE levels: IL13, IL4, IL4 receptor α, high affinity receptor for IgE, IL12b and TBET. In this particular very large study, only the polymorphisms that our group had originally reported in the gene for IL13 (Graves et al 2000) were replicated in the same direction as in our original report. No association was found with any of the other polymorphism in the seven additional genes, in spite of the fact that for many of those polymorphisms, strong associations had been originally reported in well designed studies.

What emerges from all of these studies is a very clear pattern. With few exceptions (and IL13 may indeed be one of them), associations between allergy-related phenotypes and genetic polymorphisms found in one well performed study with sufficient numbers, appropriate corrections for number of comparisons and adequately defined phenotypes, are often not reproduced in other well performed study. It is still possible that the recently published studies using a very large number of SNPs in GWA studies could provide 'hits' that are more easily reproducible and thus allow the identification of genetic variants that are strongly associated with asthma independent of context. The jury is still out, however, and it is still possible that what happened with *INSIG2* for obesity may also occur in the case of these new GWA studies. A final assessment will only be possible in the next few years, when attempts to replicate these recent 'hits' are reported in different populations.

In the meantime, currently available information clearly suggests that the prevailing paradigm to understand the genetics of complex diseases, which as explained above, supposes a more or less linear association between genetic polymorphism and phenotype, needs to be modified if genetic studies are going to be useful to understand both the pathogenesis of allergic diseases and to develop new approaches for the prevention and treatment of these diseases. Interactive effects appear to be crucial determinants of the effects of genetic polymorphisms on allergy-related phenotypes. What this means is that the polymorphisms do not have a direct relationship with the phenotype but that their effects are modulated by environmental exposures, by other polymorphisms, and by the timing during which both polymorphisms and environmental exposures interact.

References

Baldini M, Lohman IC, Halonen M, Erickson RP, Holt P, Martinez FD 1999 A polymorphism in the 5'-flanking region of the CD 14 gene is associated with circulating soluble CD14 levels and with total serum IgE. Am J Respir Cell Mol Biol 20:976–983

Braun-Fahrlander C, Riedler J, Herz U et al 2002 Environmental exposure to endotoxin and its relation to asthma in school-age children. N Engl J Med 347:869–877

Dina C, Meyre D, Samson C et al 2007 Comment on 'A common genetic variant is associated with adult and childhood obesity'. Science 315:187

Duffy DL, Martin NG, Battistutta D, Hopper JL, Mathews JD 1990 Genetics of asthma and hay fever in Australian twins. Am Rev Respir Dis 142:1351–1358

Graves PE, Kabesch M, Halonen M et al 2000 A cluster of seven tightly linked polymorphisms in the IL-13 gene is associated with total serum IgE levels in three populations of white children. J Allergy Clin Immunol 105:506–513

Herbert A, Gerry NP, McQueen MB et al 2006 A common genetic variant is associated with adult and childhood obesity. Science 312:279–283

Kedda MA, Lose F, Duffy D, Bell E, Thompson PJ, Upham J 2005 The CD14 C-159T polymorphism is not associated with asthma or asthma severity in an Australian adult population. Thorax 60:211–214

Kumar J, Sunkishala RR, Karthikeyan G, Sengupta S 2007 The common genetic variant upstream of INSIG2 gene is not associated with obesity in Indian population. Clin Genet 71:415–418

Loos RJ, Barroso I, O'Rahilly S, Wareham NJ 2007 Comment on 'A common genetic variant is associated with adult and childhood obesity'. Science 315:187

Los H, Koppelman GH, Postma DS 1999 The importance of genetic influences in asthma. Eur Respir J 14:1210–1227

Lyon HN, Emilsson V, Hinney A et al 2007 The association of a SNP upstream of INSIG2 with body mass index is reproduced in several but not all cohorts. PLoS Genet 3:e61

Maier LM, Howson JM, Walker N et al 2006 Association of IL13 with total IgE: evidence against an inverse association of atopy and diabetes. J Allergy Clin Immunol 117:1306–1313

Martinez FD 2007a Gene–environment interactions in asthma: with apologies to William of Ockham. Proc Am Thorac Soc 4:26–31

Martinez FD 2007b Genes, environments, development and asthma: a reappraisal. Eur Respir J 29:179–184

Martinez FD 2007c CD14, endotoxin, and asthma risk: actions and interactions. Proc Am Thorac Soc 4:221–225

Moffatt MF, Kabesch M, Liang L et al 2007 Genetic variants regulating ORMDL3 expression contribute to the risk of childhood asthma. Nature 448:470–473

Ober C, Hoffjan S 2006 Asthma genetics 2006: the long and winding road to gene discovery. Genes Immun 7:95–100

Poon AH, Laprise C, Lemire M et al 2004 Association of vitamin D receptor genetic variants with susceptibility to asthma and atopy. Am J Respir Crit Care Med 170:967–973

Raby BA, Lazarus R, Silverman EK et al 2004 Association of vitamin D receptor gene polymorphisms with childhood and adult asthma. Am J Respir Crit Care Med 170:1057–1065

Rosskopf D, Bornhorst A, Rimmbach C et al 2007 Comment on 'A common genetic variant is associated with adult and childhood obesity'. Science 315:187

Simpson A, John SL, Jury F et al 2006 Endotoxin exposure, CD14 and allergic disease: an interaction between genes and the environment. Am J Respir Crit Care Med 174:386–392

Smith AJ, Cooper JA, Li LK, Humphries SE 2007 INSIG2 gene polymorphism is not associated with obesity in Caucasian, Afro-Caribbean and Indian subjects. Int J Obes (Lond) 31:1753–1755

Williams LK, McPhee RA, Ownby DR et al 2006 Gene-environment interactions with CD14 C-260T and their relationship to total serum IgE levels in adults. J Allergy Clin Immunol 118:851–857

Zambelli-Weiner A, Ehrlich E, Stockton ML et al 2005 Evaluation of the CD14/-260 polymorphism and house dust endotoxin exposure in the Barbados Asthma Genetics Study. J Allergy Clin Immunol 115:1203–1209

DISCUSSION

Kleeberger: It's curious that the two Toll 4 mutations (Arbour et al 2000) are not really holding up very well in terms of prediction value for asthma or allergies. Is this fair?

Martinez: For those who don't know this, there were some significant polymorphisms/mutations in the coding region of TLR4, changing the protein. They decrease responsiveness to endotoxin in the challenge environment. The problem is that their frequency is very low, at 4–5%.

Kleeberger: Given these limitations, they still don't seem to be coming out as strongly as your CD14 polymorphisms do.

Martinez: Few people have been studying them in relation to exposure. They have been studying them as directly associated with disease. What may be happening is exactly what would happen if you do this with CD14, which is what the group in Western Australia did, and nothing is found other than heterogeneity. In our cohort study in Tucson we had 12 people who were homozygote for these TLR4 polymorphisms. What is the power to find a gene–environment interaction with 12 subjects?

Snieder: What about exposure to pets in the home? A colleague of mine in Georgia found this may have a protective effect against later development of asthma (Ownby et al 2002).

Martinez: We find significantly decreased likelihood of asthma if you have a dog in a home, but not if you have a cat. The reason for this is not that I hate cats, which I do, but most likely because cats are stealth hunters, and they have to be very clean. Dogs are collective hunters and they don't care if they smell. One could even speculate that they would like to smell, because in that way prey start running and can be caught. The amount of endotoxin in homes with dogs is higher than in houses with cats.

Kotb: you look at receptor signaling in the cells. Is this a possible mechanism?

Martinez: In the studies done among the farmers, there was significantly increased CD14 and TLR2 signaling (Ege et al 2006).

Kotb: When you have sustained endotoxin exposure, you would also expect to see sustained down-regulation of certain receptor and up-regulation of others. Are these receptors also receptors for things that mediate the allergy response?

Martinez: There are good animal studies showing that if animals are exposed to endotoxin at the same time that they are exposed to allergies, at certain doses there is a decreased likelihood of being sensitized, in other words of developing an IgE-mediated response to that allergen. Endotoxin increases the Th1-type cytokines that down regulate the Th2-type cytokines. The problem is that the dose–response curves are probably different for T and C. At low levels you are better off if you are T, because at that level the T allele is associated with more CD14 expression, which is not true at high levels.

Kleeberger: Is it really the Th1 cytokine response, or is it that the receptors are also involved in some way?

Martinez: This can be tested in animal models.

Reeve: Are there any differences among racial groups in asthma frequency?

Martinez: Yes, there are significant differences.

Reeve: How do you distinguish this from the environment?

Martinez: That's a good question. Peter LeSouef from Perth has circumnavigated the world searching for DNA from people everywhere, and has found with respect to this polymorphism an interesting latitude distribution. The C allele is frequent in the tropics but it is absent in Inuits. He has attributed the increased likelihood of having certain diseases to this distribution in relation to CD14.

Reeve: TLR2 is copy number variable. If you repeat your analysis taking this into account you may get a better dichotomy between the two groups. I can't remember whether CD14 is also copy number variable, but TLR2 is.

Uber: What I like about your paper is that it is a good example of a successful systematic research strategy that starts with heterogeneity between studies, then the biological plausibility of the environmental factor that fits the gene product. It is also an example of another crossed interaction, that some authors have suggested would be rare, but which fits in the evolutionary sense. On one hand this has public health implications. You probably wouldn't like to administer endotoxin to everyone. Because this is a crossed interaction, there is no main effect of endotoxin across the population. So are you going to look at complementary environmental factors that would be associated with asthma it the other genotype and explain why there is no main effect of G? This would be a truly personalized approach.

Martinez: Let me add that we have found very similar patterns of gene–environment interaction in still unpublished studies of polymorphisms in the promoter region of TLR2. These SNPs interact with farming environment but also with a common urban exposure, namely day care attendance, to determine allergic outcomes. We are planning to develop humanized mice for the CD14 and TLR2 genes, with the idea of determining the factors that regulate this apparent crossed or antagonistic interaction. The problem is: which cell are you going to study? We study Monomax, a monocytic cell, but is that truly the relevant cell? These are quite ancient receptors, so they are very well conserved along the way. It could be that the regulatory system is similar in mice and humans. Through the humanized approaches we were hearing about before, this could allow us to understand this a little better. The jury is still out. We don't know the exact mechanism of what we are finding is. But the fact that we are finding it with two receptors that interact with different types of agonists tells us that this may be more generalized than we have thought until now.

Uber: Have you thought of using this known causal mechanism as a screen for the other factors? You can then use whole genome data and go for a test with one degree of freedom, which is a powerful test.

Martinez: Yes. We published a study (Bieli et al 2007) done in a large population of farmers and non-farmers in Europe with respect to the use of unpasteurized milk, and found that this exposure also interacts with CD14. You talked about the possibility of exposing people to endotoxin, but exposing them in the environment may not be a good idea. The digestive system probably has better mechanisms to deal with this because it was built to deal with a huge load of bacteria. You can't give unpasteurized milk to the population because *Listeria* could kill people, so we have to think about how we can imitate nature. One of the mechanisms through which this environment may be working is that most of these families on farms give the same milk to their children without pasteurizing it.

Braithwaite: I have a question about the mechanism of the CD14 allele effects. You showed that the TT allele created a binding site for the SP1 transcription factor. Presumably there is more CD14 expressed in the TT than the CC. So that then goes on and modulates class switching, so you get a normal IgG level. But in some of your studies you showed that in the CC form there was more serum CD14?

Martinez: No, the TT has more serum CD14 in all cases, except in the experimental exposure.

Braithwaite: So the allergen is supposed to activate SP1 to bind to the TT.

Martinez: Not the allergen, but the CD14 or TLR2 agonist. That is the next step. This step probably activates a cytokine response that counteracts the Th2 response. The Th1/Th2 responses counteract each other. A strong Th1 response results in less Th2 response. At certain level of exposure to endotoxin or a TLR2 agonist there is a strong Th1 response. The hypothesis is that early life exposure to these innate immunity agonists results in an immune system that is switched away from the Th2 responses.

Braithwaite: B cells make IgGs, and at certain cell division times you go through differential class switching. That is cell cycle dependent. So what is the role of CD14 in the regulation of B cells?

Martinez: CD14 polymorphisms probably exert their effects during antigen presentation. Different types of cytokines are produced by the antigen-presenting cells that inform the T cell as to which type of immune response is preferred. In turn, this determines whether you are going to switch to IgG or IgE when the communication occurs between T cells and B cells.

Martin: Just taking Rudolph Uher's point further, are there any proposals to do trials with capsules of endotoxin for kids living in sterile cities?

Martinez: There is currently a proposal to develop in Germany unpasteurized milk that is devoid of *Listeria*. Unpasteurized milk is on sale already, but there is a huge disclaimer on this milk, because it could be contaminated with *Listeria*. The idea is to find a way in which to eliminate the *Listeria* and then use the milk in clinical trials.

Rutter: Fernando Martinez, the data you have presented provide a telling case against the people who want to divide up all genes into the good ones and the bad ones, and likewise with environments. It may be that there are some genes and environments that are awful for everyone, but in general this doesn't hold up. Numerous people have commented on and been puzzled by the apparent discrepancy between the consistent evidence from quantitative genetics that genetic influences account for a lot of variance, and the very weak effect of any of the individual identified genes. Some people suggest that this casts doubt on the twin studies, which must be peculiar in some way. The molecular genetics is telling us that genetic effects are very weak and not as strong as people thought. But another way of looking at this is that if you aren't getting main effects, but you are getting gene by environment effects, you may be going down the wrong route of needing a main effect before you look at gene × environment effects (G × E). It might be better to say that we do have some quite strong G × E effects, and this is why the twin studies appear to over-estimate the genetic effects. Is that the reason why there is a disparity?

Uber: There are two possible explanations, which are difficult to separate. It could be either G × G epistatic interactions which would be non-additive and would lead to the exaggeration of heritability in twin studies, or it could be the interaction of the genes with components of the environment that are shared by the twin pairs. The fact that epidemiological studies and secular trend studies are showing so strong effects of the environment indicates that both explanations may have some validity.

Martinez: The real difficulty is the fact that when you separate the MZ twins out, there is still the common uterus environment that you can't get rid of. In asthma, there is a significant role of intrauterine exposures. If a mother smokes during pregnancy this has a long term effect. These effects are probably related to the same family of genes that Steve was talking about before: the antioxidant/oxidant system. It is possible that when we talk about twin studies we may be seeing an interactive effect we can't detect because the environment is similar.

Martin: A potential problem with, for example, doing methylation studies on MZ twins is that what you are observing may be a zygote effect, or an effect of a particular gamete. The only way to get around that would be with inbred mouse strains. We are stuck with this. Against the importance of gene by gene interaction effects is the observation that if they are so prevalent, we would expect to see a large amount of genetic non-additivity. We would expect to see the DZ correlation way less than half the MZ correlations. Whereas for most traits it is remarkable how additive they look. This could be balancing against shared environment that we are not seeing. The only way around this is to get away from twin studies alone. With the extended twin studies we have actually done this. Lindon Eaves, Andrew Heath and I have invested huge amounts in both Virginia and Australia collecting data on 60000 individuals in the extended twin-family design. For the traits that we have looked at we see some modest departures from additivity that we wouldn't have picked with twin studies alone. It is quite remarkable for how many of the traits the twin studies are giving much the right answers.

References

Arbour NC, Lorenz E, Schutte BC et al 2000 TLR4 mutations are associated with endotoxin hyporesponsiveness in humans. Nat Genet 25:187–191

Bieli C, Eder W, Frei R et al 2007 A polymorphism in CD14 modifies the effect of farm milk consumption on allergic diseases and CD14 gene expression. J Allergy Clin Immunol 120:1308–1315

Ege MJ, Bieli C, Frei R et al 2006 Prenatal farm exposure is related to the expression of receptors of the innate immunity and to atopic sensitization in school-age children. J Allergy Clin Immunol 117:817–823

Ownby DR, Johnson CC, Peterson EL 2002 Exposure to dogs and cats in the first year of life and risk of allergic sensitization at 6 to 7 years of age. JAMA 288:963–972

13. Conclusions: taking stock and looking ahead

Michael Rutter

SGDP Centre, PO 80, Institute of Psychiatry, De Crespigny Park, Denmark Hill, London SE5 8AF, UK

It is clear that all of us accept the reality of G × E as an important biological phenomenon that has a rich potential for providing an understanding of mediating causal mechanisms for disease that will in time lead to health benefits. The notion that it is a purely statistical concept can be dismissed out of hand. On the other hand, statistical hazards bedevil the investigation of G × E, which is far from straightforward to study. There are risks of both false positives (as brought out in the chapter by Wray et al [2008, this volume] and in our discussions) and of false negatives, as shown by Uher (2008, this volume). Most especially, as emphasized throughout, there is a major need for good measures of both G and E. As Adamo and Tesson brought out (2008, this volume), the need for really good discriminating measures of the environment applies as much to the physical as to the psychosocial environment. Moreover, usually improved measurement provides more than does an increase in sample size. It is essential, too, to consider the possibility that supposed G × E actually reflects either rGE (gene–environment correlations) or G × G (i.e. synergistic combination of genes). It is crucial to have rigorous testing of both internal and external validity of the finding of G × E in any one study. We did not resolve the real difficulties that are inherent in sorting out the meaning of non-replications and partial replications. However, the consensus view seems to be that we should focus less on the specific observed G × E and more on the biological mechanisms that might be involved. There are two key reasons for that view. First, the focus has to be on the biological meaning and not the statistical interaction as such. Second, it is known that both genetic and environmental effects may be affected by contextual influences—both environmental and genetic (outside the specific G and the specific E being investigated).

Need for a range of research strategies and tactics

I was struck throughout the meeting that, across the varied medical examples, the starting point was always a known environmental influence on some specific disease outcome in which extensive heterogeneity in response had been found. Indeed, in some cases it was the failure to find a strong replicated genetic effect that led to the focus on E and then on G × E (see Martinez 2008, this volume, in relation to asthma and other atopic diseases). The prompt in the asthma example came from the observation of the apparent effect of being reared in a farm environment. Of course, it was crucial to replicate that observation in different social contexts and it was also important to have some leverage on a likely biological mediating pathway (in that case exposure to endotoxins). Similarly, the G × E focus with respect to infections began with the observation that streptococcal effects varied across the range from mild transient infectious illnesses to rapidly fatal septicemia leading to organ shutdown (see Kotb et al 2008, this volume). Of course, whilst there is no doubt about the reality of that organism causing disease, it was still necessary to check that the heterogeneity in response did not reflect variations in the pathogenicity of the organism, rather than variations in individual susceptibility.

Wray et al (2008, this volume) emphasized the value of studying discordant MZ pairs in order to test for environmental mediation of the causal effect on the disease/disorder outcome being studied. Kee-Seng Chia (2008, this volume) similarly reminded us of the utility of migration research strategies. He reported that breast cancer had been found to vary by ethnicity but migrants tended to take on the risk outcomes of the host country to which they had migrated rather than those of their original country of origin. Some kind of environmental effect was operating. There are many other types of natural experiment that can also help test for environmental mediation (see Academy of Medical Sciences 2007, Rutter 2007) and more use needs to be made of them.

Several chapters pointed to the value of intermediate phenotypes—meaning stress or challenge reactions that were involved in the environmental effect pathway. Geneticists have tended to focus on 'endophenotypes', on the basis that they are closer to the genetic effect and may, therefore, aid in the identification of susceptibility genes. The purposes here were rather different. In order to test for G × E, it is a great advantage having an E that has an immediate effect without having to wait for some disease outcome to develop. In addition, if the physiological response of the intermediate phenotype applies to other animals, and not just to humans, it provides a way of spanning animal models and the human situation. Thus, Battaglia et al (2008, this volume) described the use of response to inhalation of CO_2 (with its relationship to panic) in this way; Snieder et al (2008, this volume) reported similarly on the use of the glucose tolerance test in relation to the metabolic syndrome and cardiovascular reactivity following a stressful challenge in a standardized laboratory setting in relation to hypertension; and Adamo and Tesson (2008, this volume) pointed to the value of salt sensitivity in relation to hypertension.

All of these represent hypothesis-testing experimental medicine and some also reflected the iterative approach of going from the human situation to animal models and back again. There is sometimes a misleading tendency to assume a dichotomy between basic science in the laboratory and applied or translational medicine going from the bench to the bedside. The wrong-headed supposition is that the real scientific thinking only takes place in the former. Others have argued for the fallacy in this dichotomy—noting the two-way traffic between the two and the centrality of creative hypothesis-testing experimental medicine (see Himsworth 1962, Peters 2004). Most crucially, however, what the papers at this symposium have shown is that the same creative scientists are using both human experiments and animal models (see e.g. Battaglia et al 2008, this volume, Kleeberger & Cho 2008, this volume, and Kotb et al 2008, this volume—with the focus firmly on seeking to understand the biology).

For example, Kleeberger and Cho started with the observation in humans that oxidative stress had an adverse effect on lung function, but that bronchopulmonary dysplasia (BPD) in premature infants was less common in black babies than white ones, and was associated with HLA-A2 haplotype differences. The implication was that genetic factors might moderate susceptibility to hyperoxic lung injury. In order to investigate this possibility they turned to inbred strains of mice. Major between-strain differences in response to hyperoxia were found and a genome-wide linkage analysis followed by fine mapping of two strains identified genetic polymorphisms that were associated with differences in susceptibility. Microarray analysis demonstrated potential antioxidant defense pathways. Kleeberger and colleagues then took these animal model findings back to the human situation, providing findings on the molecular mechanisms of susceptibility to acute lung injury. The program as a whole provides a nice example of the iterative interplay of going from clinical observation to animal models in the laboratory and back to patient studies—experimental medicine at its best, exemplifying how clinical scientists can play a crucial creative role in bridging research that spans basic and clinical science and which spans animal models and human research.

Kotb's research into pathogen susceptibility provided an equally striking example of creative, hypothesis-driven bridging science. Once more, the story began with the observations in humans but then turned to animal models in order to provide stringent tests of possible mechanisms. She developed advanced recombinant inbred mouse strains to map quantitative trait loci and to sharpen biological hypotheses. The findings could then be tested in her fully humanized mouse model in which the mouse's immune system had been replaced by a human immune system. This program of research was more directly basic science in the laboratory but it reflected the same sort of translational goal, the same hypothesis-driven science, and the same focus on the identification of the basic biological pathways underlying the G × E.

In the field of genetics as a whole, there has sometimes seemed to be a tension between those advocating the need for genome-wide fishing expeditions using extremely large samples and those arguing for much more focused hypothesis-testing research strategies based on much smaller samples. We were agreed that both had a place (see Adamo & Tesson 2008, this volume, in that connection) and that it would be inappropriate to rule out one or the other. Nevertheless, as just discussed in relation to both endophenotypes and experimental medicine, hypothesis-driven studies will always be required to identify biological pathways and to chart how those may lead to the development of disease.

The literature on both medical genetics as a whole, and G × E in particular, has also involved disagreements on the relative advantages of prospective cohort studies versus case-control studies. Robertson and Poulton (2008, this volume) brought out some of the main strengths of cohort studies: the ability to provide accurate timing of environmental impacts (crucial for testing environmental mediation hypotheses); the ability to study effects on multiple disease outcomes; and the potential of examining unexpected outcomes. Their Dunedin studies of G × E well illustrate those strengths. We need to recognize, however, that for rarer outcomes, the prospective cohort studies may often need to be undertaken with high risk samples of one kind or another.

The multiple varieties of gene-environmental interplay

The focus of this meeting was strictly concerned with G × E; that is, genetic influences on susceptibility to environmental causes of disease. But the background to our discussions was provided by our awareness of the multiple other varieties of gene-environment interplay. This awareness shaped the ways in which the research had to deal with statistical hazards—such as provided by the presence of gene–environment correlations—rGE (see Rutter et al 2006). But, of course, such correlations (see Kendler & Baker 2007 for a review) also have important biological implications. The occurrence of rGE means that genes will influence the likelihood of *exposure* to risk (or protective) environments, and not just *susceptibility* to them (see e.g. Eaves et al 2003). In addition, we need to pay attention to epigenetic effects on gene expression—meaning changes that are heritable but that do not involve any change in DNA sequence (see Rutter 2006). Three key points are relevant. First, genes only have effects when they are 'expressed'. Many genes are expressed in only some body tissues and only at certain phases in development. Second, there are multiple inherited DNA elements that do not code for proteins but yet which have important effects through their influence on gene expression. We need to think about genes, not as a single entity that works (via messenger RNA) only through effects on proteins, but rather as part of a dynamic system involving multiple DNA elements. Third, environmental factors also influence gene expression—meaning that, although environments cannot change DNA sequences, they can and do influence the *effects* of genes. Moreover, it has been suggested that environmental effects on gene expression may provide a key mechanisms underlying G × E (see Abdolmaleky et al 2004).

Perhaps there are two most important messages from this symposium. First, the study of G × E should lead to a better understanding of the pathophysiological biological pathways that are involved in the origins of disease. In a sense, these can be thought of in terms of how genes get 'outside the skin' (by operating indirectly through effects on environmental exposure and environmental susceptibility); and how environments get 'under the skin' (through their effects on biology). G × E represents their bringing together. Second, it follows that the public needs to be helped to appreciate that with all the common multifactorial diseases, genes do not have deterministic effects. There are not genes 'for' any of these diseases. Not only do genes operate on proteins and not on diseases as such, but also effects are probabilistic and may often be contingent on the presence of other 'background' genes or of particular risk environments. The relevant genes are, for the most part, common allelic variations that do not have any direct pathogenic effect, rather than rare pathological mutations such as those underlying Mendelian disorders. That means that all of us must have some (probably many) genes that are associated with an increased probabilistic risk for some disease outcome. We do not necessarily develop that outcome because we do not have the required overall pattern of susceptibility genes or because we do not experience the relevant risk environment. We need to get away from simplistic assumptions about a categorical dichotomy between supposedly 'good' and 'bad' genes—with the need to get rid of the latter. The main future does not lie in gene therapy as such, but rather in interventions based on the biological understanding that will derive from the study of G × E.

Policy and practice implications

Dodge (2008, this volume) drew attention to the importance of how empirical findings and theoretical concepts are framed, because these will influence how the public responds to evidence on G × E. He noted the widespread tendency to adopt a 'blame' posture in terms of defective genes or defective environments, and the need, therefore, for scientists to emphasize the dynamic quality of the interaction effect—with its potential to be used either to improve preventive and therapeutic interventions to bring about health benefits or to foster harmful discrimination.

In several of our discussion sessions we considered this issue in relation to ethnic variations in environmental susceptibility (see e.g. Adamo & Tesson 2008, this volume, Chia 2008, this volume, Kleeberger & Cho 2008, this volume). There is no doubt that the study of such variations could be very informative in the elucidation of causal mechanisms (see Rutter & Tienda 2005). Yet the history of eugenic abuses has led many people to be very wary about genetic research. We need to help the public appreciate the positive value of genetic research but, equally, we need to be willing to accept the reality of racist prejudice and we need to be prepared to condemn such prejudice when it operates in the scientific arena. However, our overall view was that the key steps likely to be needed to bring about good G × E research using ethnic variations is to actively engage the relevant ethnic communities in that research.

We touched on the question of whether G × E findings can already influence decisions on when and how to intervene. There was a general agreement that, in time, the findings should be able to influence decisions on interventions. Most of us, however, are cautious and urged hastening slowly. That is because we need to be confident that the findings are solid, and because we need to have a better understanding of the biological mechanisms and the extent to which their effects will be shaped by other yet to be investigated genes and environments.

Implications for science

There are four main scientific implications. First, we have had some impressive examples of really good science demonstrating both how G × E can be studied effectively and that, if undertaken in the right way, can lead to major gains in scientific understanding. Second, the studies have shown not only that G × E can foster biological understanding but also it can aid the discovery of susceptibility genes. That is because if genes mainly operate through G × E they will be missed if E is ignored in genome-wide searches (see Martinez 2008, this volume). As knowledge accrues on genes affecting environmental susceptibility, a point will come when the starting point of G × E research can be based on the G, rather than just the E, as is the case now. But as gene hunters identify larger and larger numbers of susceptibility genes, the need to study G × E in order to elucidate biological mechanisms will grow, rather than fade away. Third, our deliberations have convinced all of us of the great value of bringing together scientists using different strategies to study G × E in relation to different diseases. I think that all of us go away from the meeting with useful new ideas and also a recognition of the desirability of establishing and maintaining good relationships with a range of varied scientists in our everyday scientific endeavors. Finally, it is obvious that, as demonstrated in all the presentations, collaboration is essential to solve many scientific problems. Science is a competitive enterprise but we need to engage in that enterprise in a cooperative spirit that recognizes and respects different views. That respect, of course, has to lead on to rigorous empirical studies designed to test alternative interpretations in order to gain an improved understanding of now nature 'works'. Therein lie the challenges ahead.

References

Abdolmaleky HM, Smith CL, Faraone SV et al 2004 Methylomics in psychiatry: modulation of gene–environment interactions may be through DNA methylation. Am J Med Genet B Neuropsychiatr Genet 1273:51–59

Academy of Medical Sciences 2007 Identifying the environmental causes of disease: how should we decide what to believe and when to take action? Academy of Medical Sciences, London

Adamo KB, Tesson F 2008 Gene–environment interaction and the metabolic syndrome. In: Genetic effects on environmental vulnerability to disease. Wiley Chichester (Novartis Found Symp), p 103–119

Battaglia M, Marino C, Maziade M, Molteni M, D'Amato F 2008 Gene–environment interaction and behavioural disorders: a developmental perspective based on endophenotypes. In: Genetic effects on environmental vulnerability to disease. Wiley Chichester (Novartis Found Symp), p 31–41

Chia KS 2008 Gene–environment interaction in breast cancer. In: Genetic effects on environmental vulnerability to disease, Wiley Chichester (Novartis Found Symp), p 143–150

Dodge K 2008 Practice and public policy in the era of gene–environment interactions In: Genetic effects on environmental vulnerability to disease. Wiley Chichester (Novartis Found Symp), p 87–97

Eaves L, Silberg J, Erkanli A 2003 Resolving multiple epigenetic pathways to adolescent depression. J Child Psychol Psychiatry 44:1006–1014

Himsworth H 1962 Society and the advancement of natural knowledge. Br Med J 2:1557–1563

Kendler KS, Baker JH 2007 Genetic influences on measures of the environment: a systematic review. Psychol Med 37:615–626

Kleeberger SR, Cho HY 2008 Gene by environment interactions in environmental lung diseases. In: Genetic effects on environmental vulnerability to disease. Wiley Chichester (Novartis Found Symp), p 168–178

Kotb M, Fathey N, Aziz R, Rowe S, Williams RW, Lu L 2008 Unbiased forward genetics and systems biology approaches to understanding how gene–environment interactions work to predict susceptibility and outcomes of infections. In: Genetic effects on environmental vulnerability to disease. Wiley Chichester (Novartis Found Symp), p 156–165

Martinez F 2008 Gene–environment interactions in complex diseases. In: Genetic effects on environmental vulnerability to disease. Wiley Chichester (Novartis Found Symp), p 184–192

Peters K 2004 Exceptional matters: clinical research from bedside to bench. Clin Med 4:551–566

Robertson SP, Poulton R 2008 Longitudinal studies of gene-environment interaction in common diseases—good value for money? In: Genetic effects on environmental vulnerability to disease. Wiley Chichester (Novartis Found Symp), p 128–138

Rutter M 2006 Genes and behaviour: nature–nurture interplay explained. Blackwell, Oxford

Rutter M 2007 Proceeding from observed correlation to causal inference: the use of natural experiments. Perspect Psychol Sci 2:377–395

Rutter M, Tienda M 2005 Ethnicity and causal mechanisms. Cambridge University Press, New York

Rutter M, Moffitt TE, Caspi A 2006 Gene–environment interplay and psychopathology: multiple varieties but real effects. J Child Psychol Psychiatry 47:226–261

Snieder H, Wang X, Lagou V, Penninx BWJH, Riese H, Hartman CA 2008 Role of gene–stress interactions in gene finding studies. In: Genetic effects on environmental vulnerability to disease. Wiley Chichester (Novartis Found Symp), p 71–82

Uher R 2008 Gene–environment interaction: overcoming methodological challenges. In: Genetic effects on environmental vulnerability to disease. Wiley Chichester (Novartis Found Symp), p 13–26

Wray NR, Coventry WL, James MR, Montgomery GW, Eaves LJ, Martin NG 2008 Use of monozygotic twins to investigate the relationship between 5HTTLPR genotype, depression and stressful life events: an application of Item Response Theory. In: Genetic effects on environmental vulnerability to disease. Wiley Chichester (Novartis Found Symp), p 48–59

Glossary

5HTT Serotonin transporter 5-hydroxytryptamine. The promoter region for this gene (i.e. the combination of DNA elements that influence transcription of the 5HTT gene) is referred to as 5HTTLPR.

Allele An alternative form of a gene at a particular locus—such as the A, B and O variations for the ABO blood group marker.

Allelic association Two or more neighboring loci that occur together with frequencies significantly different from those predicted from individual allele frequencies.

Animal models The use of non-human animals that have been manipulated in order to produce an organism with the same genetic mutation as one causing a particular phenotype in humans, or manipulated to manifest a form of behavior that is thought to mimic some phenotype in humans. Such models, as used in genetic research, provide an important means of investigating gene action.

Angiotensin Angiotensin is an oligopeptide in the blood that causes vasoconstriction, increased blood pressure, and release of aldosterone from the adrenal cortex.

Antioxidant Oxidative stress is implicated in many diseases and reactive oxygen molecules may be damaging. The human diet is a complex mix of oxidants and antioxidants. Antioxidants are substances that protect the body against DNA damage (molecules produced when the body breaks down food, or by environmental exposures like tobacco smoke and radiation).

Association studies Epidemiological designs based on testing whether the frequency of a specific genetic marker, or sets of markers, is higher in those affected with some disorder as compared with a control group.

Autonomic dysregulation Dysregulation of sympathetic and parasympathetic nervous system activity, which is part of the body's regulatory system for adaptation to changing environmental stimuli.

Barotrauma Physical damage to body tissues caused by a difference in pressure between an air space inside or beside the body and the surrounding gas or liquid.

Bayesian framework Bayesian Framework constitutes a statistical approach designed to quantify how new empirical evidence does, or does not, reduce uncertainty about the size of some effect. Necessarily, models have to take into account what is generally agreed with respect to reasonable assumptions.

Behavioral genetics See Quantitative genetics.

Biobank A data storage system dealing with biological measures (such as DNA).

Biomarkers Observable properties of an organism that can be used to identify exposure to an exogenous agent or to assess underlying susceptibility to disease.

Biometrics The collection, synthesis, analysis and management of quantitative data on biological communities.

BRCA BRCA 1 and 2 are genes associated with breast cancer.

Candidate gene A specific gene thought a priori to be involved in the disease process. The candidate genes can be either functional (meaning that they are involved in the postulated biological pathways leading to the disease) or positional (in that they are in genomic regions showing significant genetic linkage).

Case-control study A study that retrospectively compares a group of patients who have a medical condition with those who do not, in order to identify factors that may contribute to the cause of the condition.

C. elegans *Caenorhabditis elegans* is a free-living nematode (roundworm), about 1 mm in length, which lives in temperate soil environments. It has played a key role in genomic research.

Cell lines A permanently established cell culture that will proliferate indefinitely given appropriate fresh medium and space. This allows maintenance of DNA so that repeated samples can be analyzed without restriction to a single finite sample.

Cholinergic Related to the neurotransmitter acetylcholine.

Cohort study Study of a group of subjects sharing the same statistical or demographic characteristic, and followed over time as a group.

Confounder A variable that influences both the putative cause and its supposed consequences and thereby leads to an artefactual association.

Copy number variation (CNV) This term refers to spontaneous or *de novo* duplications that result in variations of genomic 'dosage' that could have important disease consequences.

Cronbach's alpha A statistical measure of internal consistency among the items used to develop a score.

Cohort A group of subjects sharing the same statistical or demographic characteristic, and followed over time as a group.

Comorbidity The co-occurrence of two or more disorders that are supposedly quite separate.

COMT A functional single nucleotide polymorphism (a common normal variant) of the gene for catechol-O-methyl transferase that has been shown to affect cognitive tasks broadly related to executive function, such as set shifting, response inhibition, abstract thought, and the acquisition of rules or task structure.

Congenic strains Animal genetic strains that are identical apart from at one particular locus.

Consomic   Two strains of an animal that differ by one complete chromosome pair.

Covariation   A variation in one variable that is correlated with variation in some other variable.

Dizygotic   Non-identical twins, from two zygotes.

DNA   Deoxyribonucleic acid; the double-stranded molecule, in the form of a double helix, that codes for genetic information.

DNA sequence   The order of base pairs that specifies what is inherited.

DSM-IV   *Diagnostic and statistical manual of mental disorders.* The official classification of the American Psychiatric Association; this has become quite widely used for research diagnoses throughout the world.

Dyslipidaemia   Atherogenic dyslipidemia is the overall term for blood fat disorders—high triglycerides, low HDL cholesterol and high LDL cholesterol—that foster plaque buildups in artery walls.

Ecogenetics   The interaction of genetics with the environment.

Eicosanoid   Any of a family of naturally occurring substances derived from 20-carbon polyunsaturated fatty acids; they include prostaglandins, thromboxanes, leukotrienes and epoxyeicosatrienoic acids, and function as hormones.

Endophenotype   Measurable biological characteristics thought to lie along the pathway from gene to disorder, but that are closer to the gene and are thought to have a simpler relationship with a given gene.

Endotoxin   A toxin present on the surface of Gram-negative bacteria.

Epigenetic   Changes that are heritable but that do not involve any change in DNA sequence.

Etiopathogenesis   The cause and subsequent development of an abnormal condition or of a disease.

External validity   The degree to which the results of a research study are generalizable to other populations and circumstances outside of the sample investigated.

Family-based designs   Study methods to research intergenerational or sibling traits.

Family-based genetic designs   Designs involving collection of parental DNA, as well as DNA of offspring with the disease or trait being investigated. The alleles that are not transmitted to the affected offspring are thereby used as controls in statistical analysis.

Frameshift mutations   Frameshift mutations arise by deletions and insertions that change the frame in which triplets are translated into protein.

Gene   A unit of inheritance that is made up of a stretch of DNA, more specifically a sequence of molecules called nucleotides. Inheritance 'code' encrypted in genes essentially lies in the sequence of these nucleotides.

Gene–environment correlation (rGE) Genetic influence on individual differences in exposure to particular environments.

Gene–environment interaction (G × E) Genetic influence on individual differences in susceptibility to particular environments.

Gene expression The process by which the effects of a gene have functional effects. Most genes are expressed in only some body tissues and may be expressed only at certain phases of development. Gene expression is influenced by both other genes and by environmental factors.

Genetic association studies Genetic association studies are central to efforts to identify and characterize genomic variants underlying susceptibility to multifactorial disease.

Genome The entire DNA of an organism as represented in one member of each chromosome pair.

Genome-wide study Study of the total genetic complement of an organism.

Genotype The genetic make-up of an individual; usually, however, a term restricted to the combination of alleles at a particular genetic locus.

Genome-wide association study (GWA) A study to uncover the genetic basis of a given disease, through a design that involves genotyping cases and controls at a large number of SNP markers spread (in some unspecified way) throughout the genome, and which looks for associations between the genotypes at each locus and disease status.

Haplotype A set of closely linked genetic markers that tend to be inherited together, rather than being separated during recombination.

Heterozygous Both members of a chromosome pair having different alleles at a given locus.

Heritability The proportion of variation in a particular population that is attributable to genetic influences, but note that this will include co-action with the environment. Broad heritability includes both additive and non-additive effects whereas narrow heritability concerns only additive effects.

Homozygous Both members of a chromosome pair having the same alleles at a given locus.

Hypercapnic stimulus A stimulus causing too much carbon dioxide in the blood.

ICD-10 *International classification of diseases.* This is the official classification of the World Health Organization. It has both research and clinical versions and also a number of special classifications for particular purposes.

Intermediate phenotype These are similar to endophenotypes in the assumption that they have a key role in the causal biological processes but they differ crucially in the lack of any assumption that they are necessarily closely connected to gene action.

Internal validity   The degree to which a study produces true (valid) findings within the sample investigated.

Item response theory (IRT)   This assumes a common construct underlying responses to each of a selected set of items. IRT models describe the probability of an individual endorsing a specific item as a function of the person's score on the underlying (latent) construct.

Iterative process   Repeated operations that, over a series of repetitions, provide a closer approximation to the solution of a problem.

Kinase   A type of enzyme (alternatively known as a phosphotransferase) that transfers phosphate groups from high-energy donor molecules, such as ATP, to specific target molecules (substrates); the process is termed *phosphorylation.*

Leucocytes   Generic term for white blood cells.

Ligand   An extracellular molecule that effects a change in the cytoplasm by binding to the receptor on the plasma membrane of a cell. It is used in imaging studies.

Linkage   The tendency of genes to be inherited together as a result of their location on the same chromosome, measured by percent recombination between loci.

Linkage disequilibrium   See allelic association.

Lipopolysaccharide   A molecule present in Gram-negative bacteria that contains both lipid and sugar components.

Logistic regression   A particular statistical method for determining the degree to which variation in one feature is associated with variation in another.

Macrophages   Cells within the tissues that originate from specific white blood cells called monocytes. Monocytes and macrophages are phagocytes, acting in both nonspecific defense (or innate immunity) as well as specific defense (or cell-mediated immunity) of vertebrate animals.

Mammography   X-ray of the breast to diagnose tumors by virtue of unusual areas of breast density.

MAOA   The Monoamine oxidase-A gene is involved in the production of the enzyme monoamine oxidase. This enzyme breaks down chemicals (neurotransmitters) that control mood, aggression and pleasure.

Markov chain Monte Carlo estimation   Markov chain Monte Carlo (MCMC) methods are a class of algorithms for sampling from probability distributions based on constructing a Markov chain that has the desired distribution as its equilibrium distribution. The state of the chain after a large number of steps is then used as a sample from the desired distribution. The quality of the sample improves as a function of the number of steps.

Meiosis   Cell division that occurs during gamete formation, resulting in halving the number of chromosomes so that each gamete contains only one member of each chromosome pair.

Mendelian disease The inheritance of a single gene condition that requires no particular environment for its causation and which follows particular patterns of inheritance.

Meta-analysis A technique used to combine several studies.

Metabolic syndrome The metabolic syndrome is characterized by a group of metabolic risk factors in one person, which include obesity, dyslipidaemia, elevated blood pressure, insulin resistance, prothrombotic state or pro-inflammatory state.

Methylation A chemical process that is crucially involved in epigenetic mechanisms.

Microarray A recently developed technology involving oligonucleotides or cDNA clones fixed on a glass surface for examination of hundreds to thousands of genes at the same time (sometimes referred to as a gene chip). Commonly used in a form of reverse hybridization assay to test for sequence variation in a known gene or to profile gene expression in an mRNA preparation.

Mis-sensed heterozygous mutations Mutations change a single codon and thus may cause replacement of one amino acid by another in a protein sequence.

Mitochondria The mitochondria are situated in the cytoplasm (the part of the cell outside the nucleus). Mitochondrial inheritance is entirely through the mother.

Molecular genetics Investigation of the effects of specific genes at the DNA level.

Monozygotic (MZ) Identical twins; from one zygote.

Multifactorial disorders Disorders caused by multiple genetic and environmental risk factors.

Mutation Heritable change in DNA base pair sequences.

NADPH The NADPH oxidase (nicotinamide adenine dinucleotide phosphate-oxidase) is an membrane-bound enzyme complex. It can be found in the plasma membrane as well as in the membrane of phagosome.

Nested case-control design A type of study design in which new case controls are applied into cohorts that were defined before the study begins.

Nucleotide One of the building blocks of DNA and RNA, made up of a base plus a molecule of sugar and one of phorphoric acid.

Nomological Relating to or expressing basic physical laws or rules of reasoning.

Operationalize To put something into use or operation.

Opioid A chemical substance that has a morphine-like action in the body. The main use is for pain relief. These agents work by binding to opioid receptors, which are found principally in the central nervous system and the gastrointestinal tract. The receptors in these two organ systems mediate both the beneficial effects, and the undesirable side effects.

Penetrance   The frequency with which a genotype manifests itself in a given phenotype.

PET   Positron emission tomography (PET) imaging, or a PET scan, is a diagnostic examination involving acquisition of physiologic images based on detection of radiation from the emission of positrons (tiny particles emitted from a radioactive substance administered to the patient). The subsequent images of the human body are used to evaluate a variety of diseases.

Phenotype   An observed characteristic of an individual that results from the combined effects of genotype and environment.

Phenylketonuria   An autosomal recessive metabolic disorder that involves an inability to handle the phenylalanine present in all normal human diets. If untreated, the condition results in mental retardation.

Plethysmographic measurement   Determining and registering the variations in the size or volume of an organ or limb, and hence the variations in the amount of blood in the organ or limb.

Polygenic   Polygenic inheritance of quantitative traits refers to the inheritance of a phenotypic characteristic that varies in degree and can be attributed to the interactions between two or more genes and their environment.

Polymerase chain reaction (PCR)   This is a method of greatly amplifying small amounts of DNA. It enables researchers to produce millions of copies of a specific DNA sequence in approximately two hours. This automated process bypasses the need to use bacteria for amplifying DNA.

Polymorphism   A locus with two or more alleles.

Population stratification   The situation in which alleles that are associated with ethnicity (or some other genetically influenced feature) lead to an artefactual association between some trait or disorder and the allele being studied because the cases (i.e. the individuals with the disorder) and the control group differ with respect to the allele. The association with the trait is spurious because the allelic difference arises from the genetic make-up of the population studied and not from the trait or disorder

Promoter   A combination of short sequence elements to which RNA polymerase binds in order to initiate transcription of a gene.

Proteomics   The analysis of protein expression, protein structure and protein interactions, which is fundamental for an understanding of how proteins bring about their effects and, hence, how these consequences of gene action may influence human behavior.

Quantitative (biometric) genetics   A method of studying multiple-gene influences that, together with environmental variation, result in quantitative distributions of phenotypes. Twin and adoption studies are the main strategies used.

Quantitative trait loci (QTL) Genes that contribute (along with other genes and environmental influences) to quantitative variation in some dimensional trait.

Quintile One fifth or 20% of a given amount—a term used when describing the statistical distribution of a population.

Randomized controlled trial A research design in which subjects are randomly allocated to case or control status. Randomized controlled trials are often blinded or double blinded.

Reactive oxygen species Reactive oxygen species include oxygen ions, free radicals and peroxides both inorganic and organic. They are generally very small molecules and are highly reactive due to the presence of unpaired valence shell electrons.

Recombinant inbred strains Inbred strains derived from brother-sister matings from an initial cross from two inbred progenitor strains.

RNA Ribonucleic acid.

Secular change Change over time in either the incidence of some disease/ disorder or the rate of some trait over time.

Secular trends Relatively long-term trends in a community or country.

Selection bias Systematic error that arises from the way in which subjects are included in a study.

Sensitivity analysis Statistical techniques to determine the overall strength of effects. They have been particularly employed to quantify how strong a confounder would have to be to overturn a causal inference from a case-control comparison.

Serotonergic Related to the neurotransmitter serotonin.

Single nucleotide polymorphisms (SNPs) Variations in single nucleotides that occur commonly and which are extremely useful genetic markers providing dense coverage of the genome.

Sympathovagal The concept of 'sympathovagal balance' reflects the autonomic state resulting from the coming together of sympathetic and parasympathetic influences. A variety of heart rate (HR) variability parameters have been devised as indexes of sympathovagal balance.

Syntenic Denotes the linkage of a group of genes that is found in related species.

Tetrachoric The tetrachoric correlation, for binary data, and the polychoric correlation, for ordered-category data, are excellent ways to measure rater agreement. They estimate what the correlation between raters would be if ratings were made on a continuous scale; they are, theoretically, invariant over changes in the number or 'width' of rating categories.

Trait A characteristic or phenotype, which may take the form of either a dimensional attribute or a categorical condition.

Univariate Univariate analysis separately explores each variable on its own in a data set. It looks at the range of values, as well as the central tendency of the values. It describes the pattern of response to the variable.

Wild-type The typical form of an organism, strain, gene, or characteristic as it occurs in nature.

Zygosity Whether a twin pair is monozygotic (identical) or dizygotic (fraternal).

Contributor Index

Non-participating co-authors are indicated by asterisks. Entries in bold indicate papers; other entries refer to discussion contributions.

Subject Index